Loop-like Solitons in the Theory of Nonlinear Evolution Equations

Loop-like Solitons in the Theory of Nonlinear Evolution Equations

By

V. O. Vakhnenko and E. John Parkes

Cambridge Scholars Publishing

Loop-like Solitons in the Theory of Nonlinear Evolution Equations

By V. O. Vakhnenko and E. John Parkes

This book first published 2022

Cambridge Scholars Publishing

Lady Stephenson Library, Newcastle upon Tyne, NE6 2PA, UK

British Library Cataloguing in Publication Data
A catalogue record for this book is available from the British Library

ISBN (10): 1-5275-8146-2
ISBN (13): 978-1-5275-8146-3

Contents

Preface

The physical phenomena and processes that take place in nature generally have complicated nonlinear features. This leads to nonlinear mathematical models for the real processes. There is much interest in the practical issues involved, as well as in the development of methods to investigate the associated nonlinear mathematical problems including nonlinear wave propagation. An early example of powerful mathematical methods was the development of the inverse scattering method for the Korteweg-de Vries (KdV) giving rise to the persistent interest in soliton theory as applied to many branches of science.

The modern physicist should be aware of the key aspects of nonlinear wave theory developed over the past few decades. This monograph focuses on the interconnections between a variety of different approaches and methods. The application of the theory of nonlinear evolution equations to study a new equation is always an important and sometimes rather nontrivial step. Based on our experience of the study of the Vakhnenko equation (VE), we acquaint the reader with a series of methods and approaches that can be applied to a wide class of nonlinear equations. We outline a way in which an uninitiated reader could investigate a new nonlinear equation.

Loop-like solitons are a class of interesting wave phenomena, which have been involved in some nonlinear systems. One remarkable feature of the VE is that it possesses loop-like soliton solutions.

It is our pleasure to thank and acknowledge our colleague and co-author of joint researches Dr. A.J. Morrison.

Chapter 1
Introduction

A variety of methods for examining the properties and solutions of nonlinear evolution equations are explored by using the Vakhnenko equation (VE) as an example. It is shown (Chapter 2) how the KdV equation arises in modeling the propagation of low-frequency waves in a relaxing medium. While in high-frequency cases the waves in a relaxing medium are described by an equation called now in scientific literature as the Vakhnenko equation (VE). The consideration of the VE has an interest not only from the viewpoint of the investigation of the propagation of high-frequency perturbations, but more specifically from the viewpoint of the study of methods and approaches that may be applied in the theory of nonlinear evolution equations.

By studying the VE in Chapter 3, we traced a way in which an uninitiated reader could investigate even early unknown nonlinear equations. As a first step for a new equation, it is necessary to consider the linear analogue and its dispersion relation (these steps for the equation considered here are described in Chapter 2). The next step is, where possible, to link the equation with the known nonlinear equations, as it is carried out for the VE, for example.

The solution procedure, which was used for the Vakhnenko equation (see Chapter 4), can be successfully adopted to find implicit periodic and solitary travelling-wave solutions of the Degasperis–Procesi equation in [69], the Camassa–Holm equation [71], the transformed Hirota–Satsuma-type 'shallow water wave' equation [72] and special cases thereof, namely the generalised Vakhnenko equation and the modified generalised Vakhnenko equation, the short-pulse equation [103] and other equations. An important feature of the method is that it delivers solutions in which both the dependent variable and the independent variable are given in terms of certain parameters.

In Chapter 5 the VE has been written in an alternative form, now known as the Vakhnenko-Parkes equation (VPE), by a change of independent variables. One of the main results of this chapter stated in Section 5.3 is that we have obtained a way of applying the IST method to the VPE. Keeping in mind that the IST is the most appropriate way of tackling the initial value problem, one has to formulate the associated eigenvalue problem. We have proved that the system of equations for the IST problem associated with the VPE does not contain the isospectral Schrödinger equation. Nevertheless, the analysis of the VPE in the context of the isospectral Schrödinger equation allowed us to obtain the two-soliton solution to the VPE even though, in contrast to the KdV equation, the VPE's spectral equation is not the second-order one. These results may be useful in the investigation of a new equation for which the spectral problem is unknown. Historically, once this investigation was completed, we were able to make some progress in the formulation of the IST for the VPE. In Section 7.1 it has been proved that the spectral problem associated with the VPE is of the third order.

The VPE can be written in Hirota bilinear form, as this has been carried out in Chapter 6. It is then possible to show that the VPE satisfies the 'N-soliton condition', in other words that the equation has an N-soliton solution. This solution is found by using a blend of the Hirota method and ideas originally proposed by Moloney & Hodnett. This solution is discussed in detail, including the derivation of phase shifts due to interaction between solitons. Individual solitons are hump-like. However, when transformed back into the original variables, the corresponding solution to the VE comprises N loop-like solitons. It is established that a dissipative term, with a dissipation parameter less than some limit value, does not destroy these loop-like solutions.

The Hirota method not only gives the N-soliton solution, but enables one to find a way from the Bäcklund transformation through the conservation laws and associated eigenvalue problem to the inverse scattering method. Thus the Hirota method, which can be applied only for finding solitary wave solutions or traveling wave solutions, allows us to formulate the inverse scattering method which is the most appropriate way of tackling the initial value problem (Cauchy problem). Consequently, in this case, the integrability of an equation can be regarded as proved.

Chapter 7 deals with the inverse scattering method. The inverse scattering transform (IST) method is arguably the most important discovery in the theory of solitons. The method enables one to solve the initial value problem for a nonlinear evolution equation. Moreover, it provides a proof of the complete integrability of the equation.

The idea of the inverse scattering method was first introduced for the KdV equation [94] and subsequently developed for the nonlinear Schrödinger equation [28], the mKdV equation [126, 127], the sine-Gordon equation [25, 128] and the equation of motion for a one-dimensional lattice with an exponential form of inter-site interaction (Toda lattice) [129]. It is to be remarked that the inverse scattering method is a unique theory whereby the initial value problem for the nonlinear differential equations can be solved exactly. For the KdV equation this method was expressed in general form by Lax [130].

The essence of the application of the IST is as follows. The equation of interest for study (in our case the VPE) is written as the compatibility condition for two linear equations (the Lax pair). It is significant that the spectral equation in Lax pair for the VPE is third-order. The whole Lax pair is given by Eq. (7.1.2) and Eq. (7.1.3). First, based on the ideas of Kaup and Caudrey, the initial condition $W(X, 0)$ is mapped into the spectral data $S(0)$ (the direct spectral problem). It is important that, since the variable $W(X, T)$ contained in the spectral equation evolves according to the VPE, the spectral parameter λ always retains constant values (i.e. demonstrates the isospectrality). The time evolution of the spectral data $S(T)$ is simple and linear. From a knowledge of $S(T)$ the relationships obtained by Caudrey should be invoked to reconstructed $W(X, T)$ (the inverse spectral problem). The procedure outlined allows solving the Cauchy problem for the VPE.

In Chapter 8 the standard IST method for third-order spectral problems is used to investigate solutions corresponding to bound states of the spectrum and a continuous spectrum. This leads to N-soliton solutions and M-mode periodic solutions respectively. Interactions between these types of solutions are investigated. Sufficient conditions have been proven so that the solutions become the real functions.

In Chapter 9, the standard procedure for the inverse scattering transform (IST) method is expanded for the case of multiple poles. Using the VPE as an example, we have shown how, in the IST method, to take into account the two-multiple poles, among single poles, in the discrete part of the spectral data. The special line spectrum of continuum states in the IST method, for which the mathematical procedure is similar to that for the discrete spectrum for two-multiple poles, is considered as well. In this case the account of the time-dependence is shown to be essentially different from the standard procedure. This approach can be applied to other integrable nonlinear equations.

Chapter 2
Models of relaxing medium

As a rule the behavior of media under the action of high-frequency wave perturbations is not described in the framework of equilibrium models of continuum mechanics. So, to develop physical models for wave propagation through media with complicated inner kinetics, notions based on the relaxational nature of a phenomenon are regarded to be promising. From the non-equilibrium thermodynamics standpoint, models of a relaxing medium are more general than equilibrium models. Thermodynamic equilibrium is disturbed owing to the propagation of fast perturbations. There are processes of interaction that tend to return the equilibrium. The parameters characterizing this interaction are referred to as the inner variables unlike the macro-parameters such as the pressure p, mass velocity u and density ρ. In essence, the change of macro-parameters caused by the changes of inner parameters is a relaxation process.

2.1 Evolution equation for relaxing medium

Starting from a general idea of relaxing phenomena in real media via a hydrodynamic approach, we will derive a nonlinear evolution equation for describing high-frequency waves. We restrict our attention to barotropic media. An equilibrium state equation of a barotropic medium is a one-parameter equation. As a result of relaxation, an additional variable ξ (the inner parameter) appears in the state equation

$$p = p(\rho, \xi) \tag{2.1.1}$$

and defines the completeness of the relaxation process. There are two limiting cases with corresponding sound velocities:

(i) lack of relaxation (inner interaction processes are frozen) for which $\xi = 1$:

$$p = p(\rho, 1) \equiv p_f(\rho); \tag{2.1.2}$$

(ii) relaxation is complete (there is local thermodynamic equilibrium) for which $\xi = 0$:

$$p = p(\rho, 0) \equiv p_e(\rho). \tag{2.1.3}$$

The state equations (2.1.2) and (2.1.3) are considered to be known. These relationships enable us to introduce the sound velocities for fast processes

$$c_f^2 = dp_f/d\rho; \tag{2.1.4}$$

and for slow processes

$$c_e^2 = dp_e/d\rho. \tag{2.1.5}$$

Slow and fast processes are compared using the relaxation time τ_p.

The following dynamic state equation is applied to account for the relaxation effects

$$\tau_p \left(\frac{dp}{dt} - c_f^2 \frac{d\rho}{dt} \right) + (p - p_e) = 0. \tag{2.1.6}$$

The equilibrium equations of state are considered to be known

$$\rho_e - \rho_0 = \int_{p_0}^{p} c_e^{-2} dp. \tag{2.1.7}$$

Clearly, for the fast processes ($\omega\tau_p \gg 1$) we have the relation (2.1.2), and for the slow ones ($\omega\tau_p \ll 1$) we have (2.1.3).

The substantiation of equation (2.1.6) within the framework of the thermodynamics of irreversible processes has been given in [1, 2, 3, 4]. As far as we know the first work in this field was the article by Mandelshtam and Leontovich [5] (see also Section 81 in [2]). We note that the mechanisms of the exchange processes are not defined concretely when deriving the dynamic state equation (2.1.6). In this equation the thermodynamic and kinetic parameters appear only as sound velocities c_e, c_f and relaxation time τ_p. These are very common characteristics and they can be found experimentally. Hence it is not necessary to know the inner exchange mechanism in detail. The dynamic state equation (2.1.6) enables us to take into account the exchange processes completely.

The phenomenological approach for describing the relaxation processes in hydrodynamics have been developed in many publications [2, 4, 6, 7]. The dynamic state equation was used to describe the propagation of sound in a relaxing medium [2], to take into account the exchange processes within media (gas-solid particles [4]), and to study wave fields in gas-liquid media [6] and in [7] soils. The well-known Zener phenomenological model of a standard linear solid [8] is generalized to describe the sandstone deformation [9]. Within the context of mixture theory, Biot [10] attempted to account for the non-equilibrium in velocities between components directly in the equations of motion in the form of dissipative terms. In most works, the state equation had been derived from the concept of some concrete mechanism for the inner process.

To analyze the wave motion, we use the following hydrodynamic equations in the Lagrangian coordinates:

$$\frac{\partial V}{\partial t} - \frac{1}{\rho_0}\frac{\partial u}{\partial x} = 0, \qquad \frac{\partial u}{\partial t} + \frac{1}{\rho_0}\frac{\partial p}{\partial x} = 0. \tag{2.1.8}$$

Here $V \equiv \rho^{-1}$ is the specific volume, and x is the Lagrangian space coordinate.

The closed system of equations consists of two motion equations (2.1.8) and the dynamic state equation (2.1.6). The motion equations (2.1.8) are written in Lagrangian coordinates since the state equation (2.1.6) is related to the element of mass of the medium.

Let us consider a small perturbation $p' \ll p_0$. The equations of state for fast (2.1.2) and slow (2.1.3) processes are considered to be known. They can be expanded as the power series with accuracy $O(p'^2)$

$$\begin{aligned} V_f(p_0 + p') &= V_0 - V_0^2 c_f^{-2} p' + \frac{1}{2}\left.\frac{d^2 V_f}{dp^2}\right|_{p=p_0} p'^2 + \dots\,, \\ V_e(p_0 + p') &= V_0 - V_0^2 c_e^{-2} p' + \frac{1}{2}\left.\frac{d^2 V_e}{dp^2}\right|_{p=p_0} p'^2 + \dots\,. \end{aligned} \tag{2.1.9}$$

Hereafter, the velocities c_e, c_f are related to the initial pressure p_0. Combining these two relationships with the equations of motion (2.1.8), we obtain the equation in one unknown quantity (the dash in p' is omitted) [11, 12, 13, 14]:

$$\begin{aligned}\tau_p \frac{\partial}{\partial t}\left(\frac{\partial^2 p}{\partial x^2} - c_f^{-1}\frac{\partial^2 p}{\partial t^2} + \frac{1}{2V_0^2}\left.\frac{d^2 V_f}{dp^2}\right|_{p=p_0}\frac{\partial^2 p^2}{\partial t^2}\right) \\ + \left(\frac{\partial^2 p}{\partial x^2} - c_e^{-1}\frac{\partial^2 p}{\partial t^2} + \frac{1}{2V_0^2}\left.\frac{d^2 V_e}{dp^2}\right|_{p=p_0}\frac{\partial^2 p^2}{\partial t^2}\right) = 0.\end{aligned} \tag{2.1.10}$$

A similar equation has been obtained in Ref. [1], but without nonlinear terms.

The hydrodynamic nonlinearity $p\partial p/\partial x$ and the complicated dispersive law are inherent in a medium which is described by the evolution equation (2.1.10). Now we consider the dispersive relation which follows from equation (2.1.10) after a substitution of the slow perturbation in a form $p' \sim \exp[\mathrm{i}(kx - \omega t)]$,

$$-\mathrm{i}\omega\tau_p \frac{c_e^2}{c_f^2}(\omega^2 - c_f^2 k^2) + (\omega^2 - c_e^2 k^2) = 0. \tag{2.1.11}$$

From this relationship we obtain the functional dependence $k = k(\omega)$

$$k^2 = \frac{\omega^2}{c_f^2} \cdot \frac{\tau_p^2\omega^2}{1+\tau_p^2\omega^2} \cdot \left(1 + \frac{\mathrm{i}}{\tau_p\omega} \cdot \frac{c_f^2 - c_e^2}{c_e^2} + \frac{1}{\tau_p^2\omega^2} \cdot \frac{c_e^2}{c_f^2}\right). \tag{2.1.12}$$

Taking the roots we write the result in the form $k = k' + \mathrm{i}k''$. It is clear that k'' is associated with the speed of wave attenuation as a function of the distance [2], while a value $c = \omega/k'$ can be considered as the velocity of the perturbation propagation. The expressions for k' and k'' take the form

$$k' = a_1\sqrt{\sqrt{a_2^2 + a_3^2} + a_2}, \qquad k'' = a_1\sqrt{\sqrt{a_2^2 + a_3^2} - a_2},$$

$$a_1 = \frac{\tau_p^2\omega^2}{\sqrt{2}c_f\sqrt{1+\tau_p^2\omega^2}}, \qquad a_2 = 1 + \frac{c_f^2}{\tau_p^2\omega^2 c_e^2}, \qquad a_3 = \frac{c_f^2 - c_e^2}{\tau_p\omega c_e^2}.$$

In Fig. 2.1, for example, we show the dependencies c and k'' on $\tau_p\omega$ for water-saturation soil with a concentration of air 0.1. For this

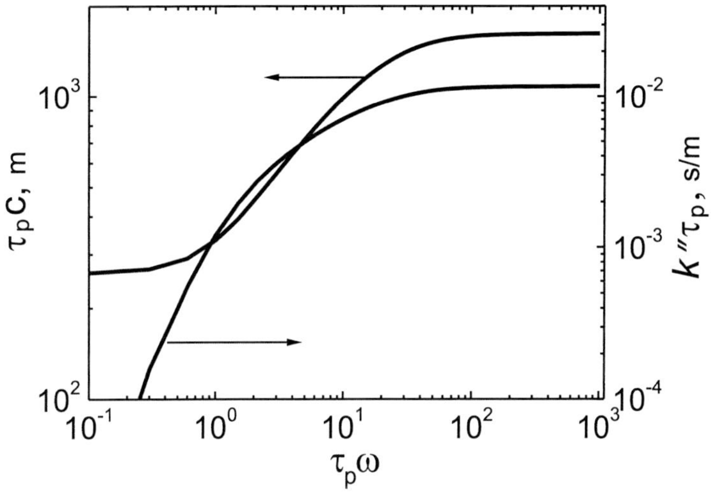

Figure 2.1: The dependencies of the velocity c and the attenuation factor k'' on frequency $\tau_p\omega$.

medium $c_f = 1620$ m/s and $c_e = 260$ m/s [7]. The velocity c increases monotonically from c_e to c_f at bottom-up sweep $\tau_p\omega$. The dependence $k'' = k''(\omega)$ indicates that at $\omega \to 0$ the dispersion is absent, while at high frequency the variable k'' becomes a constant and does not depend on ω (see Fig. 2.1) with the limit value

$$\tau_p k'' = \frac{c_f^2 - c_e^2}{2c_f^2 c_e^2}.$$

Hence, the energy in the high-frequency wave dissipates always. For this wave the pressure attenuation is the same as at fixed distance and does not depend on frequency ω.

The equation of state in the form (2.1.9) enables us to describe the effects associated with bulk viscosity of a medium. Let us show that for slow processes (since for these processes the notion of viscosity coefficient is defined, i.e. for processes in which a small deviation from equilibrium is taken into account in linear approximation) a bulk viscosity coefficient relates to the relaxation time $\tau_\rho = \tau_p c_e^2/c_\rho^2$ [1, 2, 5]

$$\zeta = \tau_\rho \rho (c_f^2 - c_e^2). \tag{2.1.13}$$

Let us rewrite (2.1.9) in a form of the power series p in $\tau_\rho d/dt$. To do it, we differentiate equation (2.1.9) with respect to time t and substitute the result into the same equation (2.1.9). Repeating several times this procedure, we obtain with required accuracy the expression

$$dp = c_e^2 d\rho + \tau_\rho (c_f^2 - c_e^2) d\dot{\rho} - \tau_\rho^2 (c_f^2 - c_e^2) d\ddot{\rho} + \dots\ . \qquad (2.1.14)$$

Let us consider two terms only in this relation. The value $c_e^2 d\rho$ associates with an increase of a pressure dp_e in infinitely slow process, i.e. $dp_e = c_e^2 d\rho$. It is noted that value p acquires more general sense than merely a pressure. With the accuracy of a sign the value $(-p)$ is nothing other than a stress π_{ii}. By definition, in the low-frequency approximation the stress is written through the bulk viscosity coefficient [2]

$$\pi_{ii} = -p_e + \zeta \frac{\partial u}{\partial x}.$$

Then it is easy to obtain the expression for the bulk viscosity coefficient in the form (2.1.13).

2.2 Low-frequency perturbations and high-frequency perturbations

Now we shall show that for low-frequency perturbations the equation (2.1.10) is reduced to the Korteweg-de Vries-Burgers (KdVB) equation, while for high-frequency waves we shall obtain the equation with hydrodynamic nonlinearity and term that appeared in the Klein-Gordon equation. To analyze the equation (2.1.10) let us apply the multiscale method [15, 16]. The value $\varepsilon \equiv \tau_p \omega$ is chosen to be small (large) parameter where the quantity ω is the characteristic frequency of wave perturbation. For the sake of convenience we rewrite the equation (2.1.10) as follows:

$$\begin{gathered} \tau_p \omega \frac{\partial}{\partial (t\omega)} \left(\frac{\partial^2 p}{\partial (x\omega)^2} - c_f^{-2} \frac{\partial^2 p}{\partial (t\omega)^2} + \alpha_f \frac{\partial^2 p^2}{\partial (t\omega)^2} \right) + \\ + \left(\frac{\partial^2 p}{\partial (x\omega)^2} - c_e^{-2} \frac{\partial^2 p}{\partial (t\omega)^2} + \alpha_e \frac{\partial^2 p^2}{\partial (t\omega)^2} \right) = 0, \end{gathered} \qquad (2.2.1)$$

$$\alpha_f = \frac{1}{2V_0^2} \frac{d^2 V_f}{dp^2}\bigg|_{p=p_0}, \qquad \alpha_e = \frac{1}{2V_0^2} \frac{d^2 V_e}{dp^2}\bigg|_{p=p_0},$$

and introduce the new independent variables [11, 12, 13, 14]

$$T_0 = t\omega, \quad X_0 = x\omega, \quad T_{-2} = t\omega/\varepsilon^2, \quad X_{-2} = x\omega/\varepsilon^2. \qquad (2.2.2)$$

The dependent variable p is a function of T_0, X_0, T_{-2}, X_{-2}, i.e. $p = p(T_0, X_0, T_{-2}, X_{-2})$. The existing derivatives in (2.2.1) are to be rewritten in the new independent variables

$$\begin{aligned}
&\frac{\partial}{\partial(x\omega)} = \frac{\partial}{\partial X_0} + \varepsilon^{-2}\frac{\partial}{\partial X_{-2}},\\
&\frac{\partial}{\partial(t\omega)} = \frac{\partial}{\partial T_0} + \varepsilon^{-2}\frac{\partial}{\partial T_{-2}},\\
&\frac{\partial^2}{\partial(x\omega)^2} = \frac{\partial^2}{\partial X_0^2} + 2\varepsilon^{-2}\frac{\partial^2}{\partial X_0 \partial X_{-2}} + \varepsilon^{-4}\frac{\partial^2}{\partial X_{-2}^2},\\
&\frac{\partial^2}{\partial(t\omega)^2} = \frac{\partial^2}{\partial T_0^2} + 2\varepsilon^{-2}\frac{\partial^2}{\partial T_0 \partial T_{-2}} + \varepsilon^{-4}\frac{\partial^2}{\partial T_{-2}^2},\\
&\frac{\partial^3}{\partial(t\omega)^3} = \frac{\partial^3}{\partial T_0^3} + 3\varepsilon^{-2}\frac{\partial^3}{\partial T_0^2 \partial T_{-2}} + 3\varepsilon^{-4}\frac{\partial^3}{\partial T_0 \partial T_{-2}^2} + \varepsilon^{-6}\frac{\partial^3}{\partial T_{-2}^3},\\
&\frac{\partial^3}{\partial(t\omega)\partial(x\omega)^2} = \frac{\partial^3}{\partial X_0^2 \partial T_0} + \varepsilon^{-2}\left(\frac{\partial^3}{\partial X_0^2 \partial T_{-2}} + 2\frac{\partial^3}{\partial T_0 \partial X_0 \partial X_{-2}}\right)\\
&\qquad + \varepsilon^{-4}\left(\frac{\partial^3}{\partial T_0 \partial X_{-2}^2} + 2\frac{\partial^3}{\partial X_0 \partial X_{-2} \partial T_{-2}}\right) + \varepsilon^{-6}\frac{\partial^3}{\partial X_{-2}^2 \partial T_{-2}}.
\end{aligned} \qquad (2.2.3)$$

It is precisely these variables that cause the equations obtained within the framework of the multiscale method [15, 16]

$$
\begin{aligned}
O(\varepsilon^{+1}) \quad &: \quad \frac{\partial}{\partial T_0}\left(\frac{\partial^2 p}{\partial X_0^2} - c_f^{-2}\frac{\partial^2 p}{\partial T_0^2} + \alpha_f \frac{\partial^2 p^2}{\partial T_0^2}\right) = 0, \\
O(\varepsilon^{0}) \quad &: \quad \frac{\partial^2 p}{\partial X_0^2} - c_e^{-2}\frac{\partial^2 p}{\partial T_0^2} + \alpha_e \frac{\partial^2 p^2}{\partial T_0^2} = 0, \\
O(\varepsilon^{-1}) \quad &: \quad \left(\frac{\partial^3}{\partial X_0^2 \partial T_{-2}} + 2\frac{\partial^3}{\partial T_0 \partial X_0 \partial X_{-2}}\right) p \\
&\qquad -3c_f^{-2}\frac{\partial^3 p}{\partial T_0^2 \partial T_{-2}} + 3\alpha_f \frac{\partial^3 p^2}{\partial T_0^2 \partial T_{-2}} = 0, \\
O(\varepsilon^{-2}) \quad &: \quad \frac{\partial^2 p}{\partial X_0 \partial X_{-2}} - c_e^{-2}\frac{\partial^2 p}{\partial T_0 \partial T_{-2}} + \alpha_e \frac{\partial^2 p^2}{\partial T_0 \partial T_{-2}} = 0, \qquad (2.2.4) \\
O(\varepsilon^{-3}) \quad &: \quad \left(\frac{\partial^3}{\partial T_0 \partial X_{-2}^2} + 2\frac{\partial^3}{\partial X_0 \partial X_{-2} \partial T_{-2}}\right) p \\
&\qquad -3c_f^{-2}\frac{\partial^3 p}{\partial T_0 \partial T_{-2}^2} + 3\alpha_f \frac{\partial^3 p^2}{\partial T_0 \partial T_{-2}^2} = 0, \\
O(\varepsilon^{-4}) \quad &: \quad \frac{\partial^2 p}{\partial X_{-2}^2} - c_e^{-2}\frac{\partial^2 p}{\partial T_{-2}^2} + \alpha_e \frac{\partial^2 p^2}{\partial T_{-2}^2} = 0, \\
O(\varepsilon^{-5}) \quad &: \quad \frac{\partial}{\partial T_{-2}}\left(\frac{\partial^2 p}{\partial X_{-2}^2} - c_f^{-2}\frac{\partial^2 p}{\partial T_{-2}^2} + \alpha_f \frac{\partial^2 p^2}{\partial T_{-2}^2}\right) = 0,
\end{aligned}
$$

to be partially uncoupled [17, 18, 19, 20, 21, 22]. The two leading equations depend on T_0 and X_0 only, while the last two equations include the independent variables T_{-2} and X_{-2} only. Thus, the low-frequency perturbations are described by the two leading equations, and the high-frequency perturbations by the last two equations. An interaction between these perturbations is described by the three center equations. A similar approach was applied to obtain the evolution equation with cubic nonlinearity [23, 24].

Let us write out the equations of motion for low-frequency and high-frequency perturbations in the initial variables x and t. For low-frequency perturbations the main terms $\partial^2 p/\partial X_0^2$ and $c_e^{-2}\partial^2 p/\partial T_0^2$ (and only they) appear in the first and second equations of the system (2.2.4), while for high-frequency perturbations the main terms

$\partial^2 p/\partial X_{-2}^2$ and $c_f^{-2}\partial^2 p/\partial T_{-2}^2$ (and only they) appear in the sixth and seventh equations of the system (2.2.4).

For low-frequency perturbations ($\tau_p\omega \ll 1$) propagating in one direction ($\partial/\partial x - c_e^{-1}\partial/\partial t \simeq 2\partial/\partial x$), we obtain an evolution equation

$$\frac{\partial p}{\partial t} + c_e\frac{\partial p}{\partial x} + \alpha_e c_e^3 p\frac{\partial p}{\partial x} - \beta_e\frac{\partial^2 p}{\partial x^2} + \gamma_e\frac{\partial^3 p}{\partial x^3} = 0, \tag{2.2.5}$$

$$\alpha_e = \frac{1}{2V_0^2}\left.\frac{d^2 V_e}{dp^2}\right|_{p=p_0}, \qquad \beta_e = \frac{c_e^2\tau_p}{2c_f^2}(c_f^2 - c_e^2),$$

$$\gamma_e = \frac{c_e^3\tau_p^2}{8c_f^4}(c_f^2 - c_e^2)(c_f^2 - 5c_e^2).$$

This equation can be derived in the following way. A dispersion relation for the linearized equation (2.1.10) can be written down with an accuracy $O(k^3)$ in the form $\omega = c_e k + \mathrm{i}\beta_e k^2 - \gamma_e k^3$, if the terms $\partial p/\partial x$ and $c_e^{-1}\partial p/\partial t$ are the main ones. For this dispersion relation we write a linear equation in which a nonlinear term is reconstructed in agreement with the initial equation.

The equation (2.2.5) is the well-known KdVB equation. It is encountered in many areas of physics to describe nonlinear wave processes [25, 26, 27, 28, 29]. In [30] it was shown how hydrodynamic equations reduce to either the KdV or Burgers equation according to the choices for the state equation and the generalized force when analyzing the gasdynamical waves, waves in shallow water [30], hydrodynamic waves in cold plasma [31], and ion-acoustic waves in cold plasma [32].

As is known, the investigation of the KdV equation ($\beta_e = 0$) in conjunction with the nonlinear Schrödinger (NLS) and sine-Gordon equations gives rise to the theory of solitons [25, 27, 28, 29, 30, 33, 34, 35, 36, 37]. As well as having soliton solutions, these equations have other inherent striking properties, in particular integrability. The equations can be integrated, for example, by the inverse scattering method. Details on the study of the aforementioned equations can be found in the monographs [25, 27, 28]. In general, the existence of soliton solutions to a nonlinear evolution equation points to distinctive features for the equation such as integrability, the applicability of the inverse scattering method, the Hirota method and Bäcklund transformation, and the existence of conservation laws. Consequently, the finding of soliton solutions for a new evolution equation is of considerable interest.

For high-frequency perturbations ($\tau_p\omega \gg 1$), using the last two equations of the system (2.2.4), we get the following evolution equation:

$$\frac{\partial^2 p}{\partial x^2} - c_f^{-2}\frac{\partial^2 p}{\partial t^2} + \alpha_f c_f^2 \frac{\partial^2 p^2}{\partial x^2} + \beta_f \frac{\partial p}{\partial x} + \gamma_f p = 0. \tag{2.2.6}$$

$$\alpha_f = \frac{1}{2V_0^2}\left.\frac{d^2 V_f}{dp^2}\right|_{p=p_0}, \quad \beta_f = \frac{c_f^2 - c_e^2}{\tau_p c_e^2 c_f}, \quad \gamma_f = \frac{c_f^4 - c_e^4}{2\tau_p^2 c_e^4 c_f^2}.$$

In addition to the nonlinear term with coefficient α_f, the equation has dissipative $\beta_f \partial p/\partial x$ and dispersive $\gamma_f p$ terms. If $\alpha_f = \beta_f = 0$, this is a linear Klein-Gordon equation. There is a Green function for this equation [38] that enables us to find the solution in quadrature, at least. The numerical solutions of the Klein-Gordon equation modeling the propagation of high-frequency perturbations in gas-liquid media have been presented in [39]. A similar evolution equation for high-frequency perturbations was described in a monograph by Whitham [40]. However, it coincides with Eq. (2.2.6) only when $\alpha_f = 0$ and $\gamma_f = 0$.

Landau and Lifshitz showed that for high frequencies the dissipative term under high transport of heat agrees with the corresponding term in the equation (2.2.6) (see section 79 and 81 in [2]). Thus, the dynamic state equation (2.1.9) enables us to take into account the dissipative processes completely. But the form of the dissipative terms describing the inner exchange processes (transport of heat and momentum) are different for the high and low frequencies.

We call attention to the fact that the dispersion relations $\omega = \omega(k)$ for the linearized equations (2.2.5) and (2.2.6) have been restricted by the finite power series in k and in k^{-1}, respectively:

$$\omega = c_e k + \mathrm{i}\beta_e k^2 - \gamma_e k^3, \qquad \tau_p\omega \ll 1,$$

$$\omega^2 = c_f^2 k^2 (1 + \mathrm{i}\beta_f k^{-1} - \gamma_f k^{-2}), \qquad \tau_p\omega \gg 1.$$

At the time we were carrying out our research, it turned out that equation (2.2.6) had not been investigated much. This is likely connected with the fact, noted by Whitham in Ref. [40], that high-frequency perturbations attenuate very quickly. However in Whitham's monograph [40], the evolution equation (2.2.6) without nonlinear and dispersive terms was considered. Certainly, the lack of such terms restricts the class of solutions. At least, there is no solution in the form of a solitary wave which is caused by nonlinearity and dispersion.

2.3 Evolution equation for high-frequency perturbations

The equation (2.2.6), which we are interested in,

$$\frac{\partial^2 p}{\partial x^2} - c_f^{-2}\frac{\partial^2 p}{\partial t^2} + \alpha_f c_f^2 \frac{\partial^2 p^2}{\partial x^2} + \beta_f \frac{\partial p}{\partial x} + \gamma_f p = 0$$

is written down in a dimensionless form. Let us restrict our consideration to the propagation of high-frequency waves in positive direction x, then with necessary accuracy we can write the operator

$$\frac{\partial^2}{\partial x^2} - c_f^{-2}\frac{\partial^2}{\partial t^2} =$$

$$= \left(\frac{\partial}{\partial x} - c_f^{-1}\frac{\partial}{\partial t}\right)\left(\frac{\partial}{\partial x} + c_f^{-1}\frac{\partial}{\partial t}\right) \to 2\frac{\partial}{\partial x}\left(\frac{\partial}{\partial x} + c_f^{-1}\frac{\partial}{\partial t}\right).$$

In the moving coordinates system with velocity c_f, the equation has the form in dimensionless variables

$$\tilde{x} = \sqrt{\frac{\gamma_f}{2}}(x - c_f t), \quad \tilde{t} = \sqrt{\frac{\gamma_f}{2}}c_f t, \quad \tilde{u} = \alpha_f c_f^2 p$$

(tilde over variables $\tilde{x}, \tilde{t}, \tilde{u}$ is omitted) [11, 12, 13, 41]

$$\frac{\partial}{\partial x}\left(\frac{\partial}{\partial t} + u\frac{\partial}{\partial x}\right)u + \alpha\frac{\partial u}{\partial x} + u = 0. \qquad (2.3.1)$$

The constant $\alpha = \beta_f/\sqrt{2\gamma_f}$ is always positive. The equation (2.3.1) without the dissipative term has the form of the nonlinear equation [41, 42]

$$\frac{\partial}{\partial x}\left(\frac{\partial}{\partial t} + u\frac{\partial}{\partial x}\right)u + u = 0. \qquad (2.3.2)$$

Historically, the equation (2.3.2) has been called the Vakhnenko equation (VE) and we shall use this name subsequently.

It is interesting to note that equation (2.3.2) follows as a particular limit of the following generalized Korteweg-de Vries equation

$$\frac{\partial}{\partial x}\left(\frac{\partial u}{\partial t} + u\frac{\partial u}{\partial x} - \beta\frac{\partial^3 u}{\partial x^3}\right) = \gamma u \qquad (2.3.3)$$

derived by Ostrovsky [43] to model small-amplitude long waves in a rotating fluid (γu is induced by the Coriolis force) of finite depth. Subsequently, equation (2.3.2) was known by different names in the literature, such as the Ostrovsky-Hunter equation, the short-wave equation, the reduced Ostrovsky equation and the Ostrovsky-Vakhnenko equation depending on the physical context in which it is studied.

The consideration here of equation (2.3.2) has an interest not only from the viewpoint of the investigation of the propagation of high-frequency perturbations, but more specifically from the viewpoint of the study of methods and approaches that may be applied in the theory of nonlinear evolution equations.

Chapter 3
The travelling-wave solutions

By investigating equation (2.3.2), we will trace a way in which an uninitiated reader could investigate a new nonlinear equation. As a first step for a new equation, it is necessary to consider the linear analogue and its dispersion relation (these steps for equations (2.2.5) and (2.2.6) are described already in Chapter 2). The next step is, where possible, to link the equation with a known nonlinear equation.

3.1 The connection of the VE with the Whitham equation

Now we show how an evolution equation with hydrodynamic nonlinearity can be rewritten in the form of the Whitham equation. The general form of the Whitham equation is as follows [40]:

$$\frac{\partial u}{\partial t} + u\frac{\partial u}{\partial x} + \int_{-\infty}^{\infty} K(x-s)\frac{\partial u}{\partial s}ds = 0. \tag{3.1.1}$$

On the one hand, this equation (3.1.1) has the nonlinearity of hydrodynamic type; on the other hand, it is known (see, Section 13.14 in [40]) that the kernel $K(x)$ can be selected to give the dispersion required. Indeed, the dispersion relation $c(k) = \omega(k)/k$ and the kernel $K(x)$ are connected by means of the Fourier transformation

$$c(k) = F[K(x)], \qquad K(x) = F^{-1}[c(k)]. \tag{3.1.2}$$

Consequently, for the dispersion relation $\omega = -1/k$ corresponding to the linearized version of (2.3.2), the kernel is as follows

$$K(x) = F^{-1}[-1/k^2] = \tfrac{1}{2}|x|. \tag{3.1.3}$$

Thus, the VE (2.3.2) is related to the particular Whitham equation [40]

$$\frac{\partial u}{\partial t} + u\frac{\partial u}{\partial x} + \frac{1}{2}\int_{-\infty}^{\infty} |x-s|\frac{\partial u}{\partial s}ds = 0. \tag{3.1.4}$$

Since we can reduce the VE to the Whitham equation, we can assert that the VE shares interesting properties with the Whitham equation; in particular, it describes solitary wave-type formations, have periodic solutions and explains the existence of the limiting amplitude [40]. An important property is the presence of conservation laws for waves decreasing rapidly at infinity, namely

$$\frac{d}{dt}\int_{-\infty}^{\infty} u dx = 0, \quad \frac{d}{dt}\int_{-\infty}^{\infty} u^2 dx = 0, \quad \frac{d}{dt}\int_{-\infty}^{\infty} \left(\tfrac{1}{3}u^3 + \widehat{K}u\right) dx = 0, \tag{3.1.5}$$

where by definition $\widehat{K}u = \int_{-\infty}^{\infty} K(x-s)u(s,t)ds$.

For equation (2.3.1) the kernel is $K(x) = \frac{1}{2}[\alpha(2\Theta(x) - 1) + |x|]$, where $\Theta(x)$ is the Heaviside function. Hence, (2.3.1) can be written down as

$$\frac{\partial u}{\partial t} + u\frac{\partial u}{\partial x} + \alpha u + \frac{1}{2}\int_{-\infty}^{\infty} |x-s|\frac{\partial u}{\partial s}ds = 0. \tag{3.1.6}$$

It is important that there is no derivative in the dissipative term αu of Eq. (3.1.6).

3.2 Loop-like stationary solutions of the VE

An important step in the investigation of nonlinear evolution equations is to find travelling-wave solutions. These are solutions that are stationary with respect to a moving frame of reference. In this case, the evolution equation (a partial differential equation) becomes an ordinary differential equation (ODE) which is considerably easier to solve.

For the VE (2.3.2) it is convenient to introduce a new dependent variable z and new independent variables η and τ defined by

$$z = (u - v)/|v|, \quad \eta = (x - vt - x_0)/|v|^{1/2}, \quad \tau = t|v|^{1/2}, \tag{3.2.1}$$

where v and x_0 are arbitrary constants, and $v \neq 0$. Then the VE becomes

$$z_{\eta\tau} + (zz_\eta)_\eta + z + c = 0, \text{ where } c := \frac{v}{|v|}. \tag{3.2.2}$$

$c = \pm 1$ corresponding to whether $v \gtrless 0$. We now seek stationary solutions of (3.2.2) for which z is a function of η only so that $z_\tau = 0$ and z satisfies the ODE

$$(zz_\eta)_\eta + z + c = 0, \tag{3.2.3}$$

After one integration (3.2.3) gives

$$\begin{aligned} &\tfrac{1}{2}(zz_\eta)^2 = f(z), \\ &f(z) = -\tfrac{1}{3}z^3 - \tfrac{1}{2}cz^2 + \tfrac{1}{6}c^3 A = -\tfrac{1}{3}(z - z_1)(z - z_2)(z - z_3), \end{aligned} \tag{3.2.4}$$

where A is a constant. It is easy to verify that if there are complex roots, the value z tends to minus infinity, and this contradicts the physical statement of the problem. Indeed, if we have only one real root, the graph of the function $f(z)$ (see fig. 3.1) crosses the Oz axis once. Thus as $z \to +\infty$ we have $f \to -\infty$ and as $z \to -\infty$ we have $f \to +\infty$. But since the trinomial in (3.2.4) should always be positive in the integration region, as follows from the l.h.s. of (3.2.4), this region extends in z from minus infinity to the value of the single real root. This means the perturbation amplitude $u = (z + c)v$ also tends to minus infinity, which does not correspond to the physical statement of the problem. So, all roots of the trinomial should be real. This requires that $0 \leq A \leq 1$. Note that there are turning points at $z = 0$ and $z = -c$. For periodic or solitary-wave solutions, z_1, z_2 and z_3 are real constants. For definiteness we shall assume that $z_1 \leq z_2 \leq z_3$. Three ways of calculating the roots are given in the Appendix in Section 3.4. From (A.7), we can deduce that for $v > 0$ always the root $z_3 \in [0, 0.5]$ as indicated by curve 2 in Fig. 3.1(a); curve 1 corresponds to $A = 1$ and curve 3 corresponds to $A = 0$. Similarly, for $v < 0$ always $z_3 \in [1, 1.5]$ as indicated by curve 2 in

Fig. 3.1(b); curve 1 corresponds to $A = 0$ and curve 3 corresponds to $A = 1$. It also follows from (A.7) that always $z_1 < 0$, but $z_2 < 0$ for $v > 0$ and $z_2 > 0$ for $v < 0$. Thus the nature of the solutions depends on the sign of v. However, the integration of (3.2.4) is not affected by the sign of v.

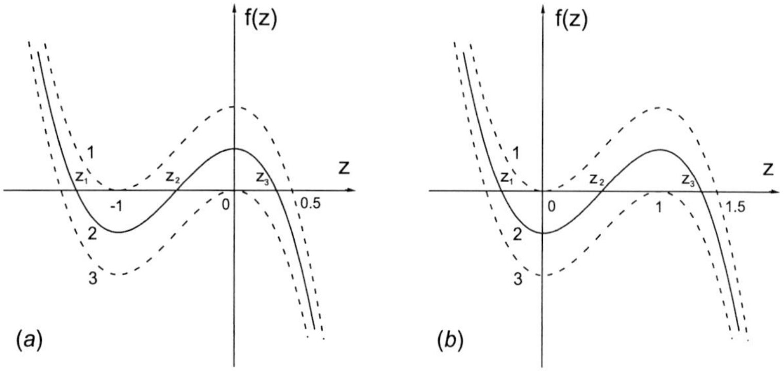

Figure 3.1: The graph of the trinomial $f(z)$: (a) $v > 0$, (b) $v < 0$. The integration region is the interval (z_2, z_3).

The integration region of (3.2.4) is the interval (z_2, z_3) where $f(z) > 0$ (see fig. 3.1). At the points $z = z_2$ and $z = z_3$ the derivatives z_η are zero. Hence, we have the relation

$$\pm\sqrt{\tfrac{2}{3}}\eta = \int_{z}^{z_3} \frac{zdz}{\sqrt{(z-z_1)(z-z_2)(z_3-z)}}. \tag{3.2.5}$$

On using results 236.00 and 236.01 of [44], we may integrate (3.2.5) to obtain

$$\eta = \frac{1}{p}[z_1 F(\varphi, k) + (z_3 - z_1)E(\varphi, k)], \tag{3.2.6}$$

where

$$\sin^2\varphi = \frac{z_3 - z}{z_3 - z_2}, \quad k^2 = \frac{z_3 - z_2}{z_3 - z_1}, \quad p^2 = \frac{(z_3 - z_1)}{6}. \tag{3.2.7}$$

In the notation of [44], $F(\varphi, k)$ and $E(\varphi, k)$ are incomplete elliptic integrals of the first and second kind respectively. We have chosen the constant of integration in (3.2.6) to be zero so that $z = z_3$ at $\eta = 0$. The relations (3.2.6) and (3.2.7) give the required solution in parametric form, with z and η as functions of the parameter φ.

An alternative route to the solution is to follow the procedure described in [45]. We introduce a new independent variable ζ defined by

$$\frac{d\eta}{d\zeta} = z \tag{3.2.8}$$

so that (3.2.4) becomes

$$\tfrac{1}{2} z_\zeta^2 = f(z). \tag{3.2.9}$$

By means of result 236.00 of [44], (3.2.9) may be integrated to give $w = F\,(\varphi|m)$, where $m{:} = k^2$ and $w = p\zeta$. Here we have used the notation of Chapter 17 in [110]. Thus, on noting that $\sin\varphi = \mathrm{sn}(w|m)$, where sn is a Jacobian elliptic function, we have

$$z = z_3 - (z_3 - z_2)\,\mathrm{sn}^2(w|m). \tag{3.2.10}$$

With result 310.02 of [44], (3.2.8) and (3.2.10) give

$$\eta = \frac{1}{p}[z_1 w + (z_3 - z_1)E(w|m)], \tag{3.2.11}$$

where $E(w|m)$ is the incomplete elliptic integral of the second kind (in the notaion of [110]). Relations (3.2.10) and (3.2.11) are equivalent to (3.2.7) and (3.2.6) respectively and give the solution in parametric form with z and η in terms of the parameter w.

We define the wavelength λ of the solution as the amount by which η increases when φ increases by π, or equivalently when w increases by $2K(m)$, where $K(m)$ is the complete elliptic integral of the first kind. It follows from (3.2.11) that

$$\lambda = \frac{2}{p}|z_1 K(m) + (z_3 - z_1)E(m)|, \tag{3.2.12}$$

where $E(m)$ is the complete elliptic integral of the second kind.

As mentioned previously, the VE has two families of solutions corresponding to $v > 0$ and $v < 0$, respectively. We now describe these two cases in detail.

With $v > 0$ we have $c = 1$. Then, with $0 < A < 1$, there are periodic loop solutions with $0 < m < 1$, $z_2 \in (-1, 0)$ and $z_3 \in (0, 0.5)$; an example of such a periodic wave is illustrated by curve 2 in Fig. 3.2. The loop-like nature of these periodic waves is due to the fact that $z = 0$ is in the interval (z_2, z_3). For small z, (3.2.4) gives $z_\eta \simeq \pm\sqrt{A/(3z)}$ so that $|z_\eta| \to \infty$ as $z \to 0$. It follows that, when $z = 0$, the solution curve is normal to the η axis. $A = 1$ gives the solitary wave limit for which $z_1 = z_2 = -1$ and $z_3 = 1/2$ so that $m = 1$. As $\mathrm{sn}(w|1) \equiv \tanh w$ and $E(w|1) \equiv \tanh w$, (3.2.10) and (3.2.11) reduce to

$$z = \tfrac{1}{2} - \tfrac{3}{2}\tanh^2 w, \qquad \eta = -2w + 3\tanh w, \tag{3.2.13}$$

where $w = \zeta/2$. In terms of the original dependent variable u and the new independent variable ζ, (3.2.13) gives

$$u = \tfrac{3}{2} v \,\mathrm{sech}^2(\zeta/2), \qquad \eta = -\zeta + 3\tanh(\zeta/2) \tag{3.2.14}$$

as illustrated by curve 1 in Fig. 3.2. When $u/v = 1$, this solution curve is normal to the η axis at the points $\eta = \mp W/2$, where W is the maximum width of the loop. On putting $z = 0$ and $\eta = \mp W/2$ into (3.2.13), we find that

$$W = 2\sqrt{3} - 4\tanh^{-1}\left(\frac{1}{\sqrt{3}}\right). \tag{3.2.15}$$

With $v < 0$ we have $c = -1$. Then, with $0 < A < 1$, there are periodic solutions with $0 < m < 1$, $z_2 \in (0,\ 1)$ and $z_3 \in (1,\ 1.5)$; an example of such a periodic wave is illustrated by curve 2 in Fig. 3.3. With $A = 0$, $z_1 = z_2 = 0$ and $z_3 = 3/2$ so that $m = 1$ and $\lambda = 6$. (3.2.10) and (3.2.11) reduce to

$$z = \tfrac{3}{2} - \tfrac{3}{2}\tanh^2 w, \qquad \eta = 3\tanh w, \tag{3.2.16}$$

where $w = \zeta/2$. Elimination of $\tanh w$ now gives

$$z = \tfrac{3}{2} - \tfrac{1}{6}\eta^2, \qquad -3 \le \eta \le 3 \tag{3.2.17}$$

from which we can construct a weak solution in the form of a spatially periodic inverted 'paraboidal' wave (i.e. a corner wave) of amplitude 3/2 given by

$$z = z(\eta - 6j), \quad -3 \le \eta - 6j \le 3, \quad j = 0, \pm 1, \pm 2, \ldots. \tag{3.2.18}$$

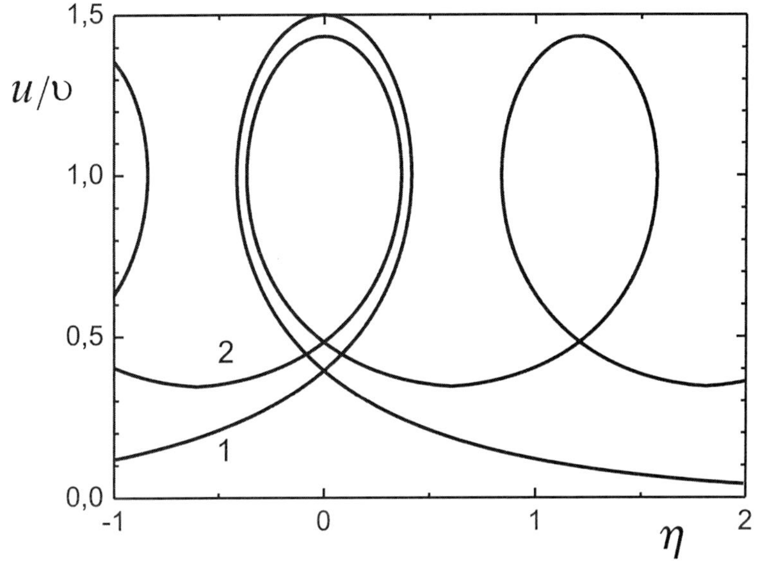

Figure 3.2: Travelling wave solutions with $v > 0$.

In terms of the original dependent variable u, (3.2.17) and (3.2.18) become

$$u/|v| = \tfrac{1}{2} - \tfrac{1}{6}\eta^2, \qquad -3 \le \eta \le 3 \tag{3.2.19}$$

and

$$u/|v| = u(\eta - 6j)/|v|, \quad -3 \le \eta - 6j \le 3, \quad j = 0, \pm 1, \pm 2, \ldots \tag{3.2.20}$$

respectively. The latter is shown by the curve 1 in Fig. 3.3. For $A \simeq 1$ the solution has a sinusoidal form (see curve 3 in Fig. 3.3). Note that there are no solitary wave solutions with $v < 0$. This is due to the fact that $z = 0$ is not in the interval (z_2, z_3).

A remarkable feature of the equation (2.3.2) is that it has a solitary wave (3.2.14) which has a loop-like form, i.e. it is a multi-valued function (see Fig. 3.2). Whilst loop solitary waves (3.2.14) are rather intriguing, it is the solution to the initial value problem that is of more interest in a physical context. An important question is the stability of the loop-like solutions. Although the analysis of stability does not link with the theory of solitons directly, however, the method ap-

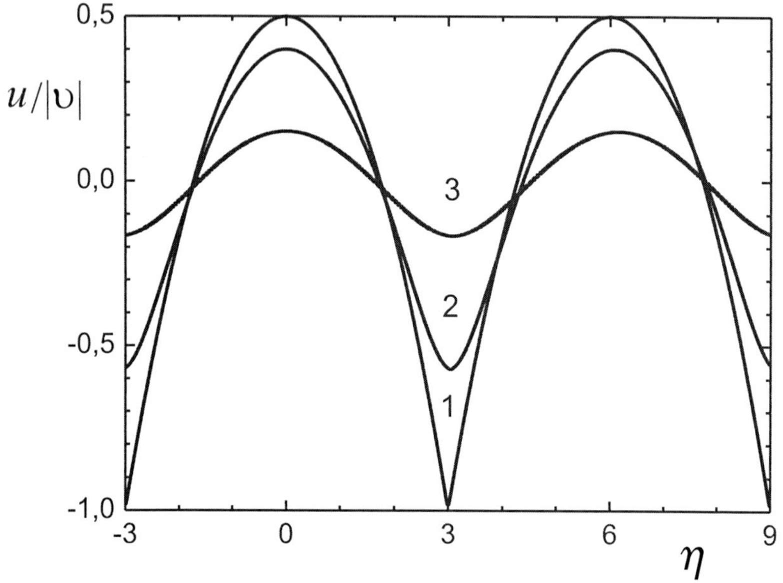

Figure 3.3: Travelling wave solutions with $v < 0$.

plied in Section 3.3.1 is instructive, since it is successful in a nonlinear approximation.

We note that the notion of a 'soliton' will be defined later. We will prove (see Section 6.5) that the solitary wave (3.2.14) is, in fact, a soliton. Here we point out only that the soliton is a local travelling wave pulse with remarkable stability and particle-like properties.

3.3 Stability, ambiguity and interpretation of the loop-like solutions

3.3.1 Stability

From a physical viewpoint, the stability or otherwise of solutions is essential to their interpretation. Some methods for the investigation of the stability of nonlinear waves were discussed by Infeld and Rowlands in Chapter 8 of [46] and references therein. One such method is the so-called k-expansion method. It is restricted to long wavelength perturbations of small amplitude. It has been applied successfully to a variety of generic nonlinear evolution equations (see [47] for example) and specific physical systems (see [48] for example). A particularly informative description of the method is given in [49] in the context of the Zakharov-Kuznetsov equation. Some criticism was leveled at the work in [49] by Das *et al* [50]; however, after a detailed reinvestigation of the problem, Das *et al* [50] vindicated the method used in [49].

The k-expansion method was applied to the VE (2.3.2) in [42] and is outlined as follows. We assume a perturbed solution of (3.2.4) in the form

$$z = z_0(\eta) + \{\delta z(\eta) \exp[\mathrm{i}(k\eta - \omega\tau)] + \mathrm{cc}\}, \tag{3.3.1}$$

where z_0 is the periodic solution given by (3.2.10) and (3.2.11), $\delta z(\eta)$ is a complex function with period λ given by (3.2.12)), k is a real constant, ω is a constant (possibly complex), and cc denotes complex conjugate. Substitution of (3.3.1) into (3.2.2) and linearization with respect to δz yields

$$\mathcal{L}\delta z = f, \tag{3.3.2}$$

where the linear operator $\mathcal{L}$ and f are given by

$$\mathcal{L}\delta z\colon = (z_0\delta z)_{\eta\eta} + \delta z, \qquad f\colon = (-\omega k + k^2 z_0)\delta z + \mathrm{i}[\omega\delta z_\eta - 2k(z_0\delta z)_\eta],$$

respectively. As (3.2.3) implies that $\mathcal{L}z_{0\eta} = 0$, we may deduce that, for (3.3.2) to have periodic solutions, the condition

$$\langle z_0 z_{0\eta} f\rangle = 0 \tag{3.3.3}$$

must be satisfied, where $\langle\cdot\rangle$ denotes an integration over the wavelength λ.

Formally, the solution of (3.3.2) is

$$\delta z = z_{0\eta}\vartheta, \tag{3.3.4}$$

where

$$\vartheta_\eta = \left(D + \int z_0 z_{0\eta} f \, d\eta\right) / (z_0 z_{0\eta})^2 \tag{3.3.5}$$

and D is a constant determined from

$$\left\langle \left(D + \int z_0 z_{0\eta} f \, d\eta\right) / (z_0 z_{0\eta})^2 \right\rangle = 0. \tag{3.3.6}$$

As δz appears on the right-hand side of (3.3.4) via f, we solve (3.3.4) iteratively by assuming that k is small in comparison with $2\pi/\lambda$ (so that the perturbations in (3.3.1) have long wavelength) and introduce the expansions

$$\delta z = \delta z_0 + k\delta z_1 + \cdots , \qquad \omega = k\omega_1 + k^2\omega_2 + \cdots ,$$

so that

$$f = kf_1 + k^2 f_2 + \cdots , \quad \vartheta = \vartheta_0 + k\vartheta_1 + \cdots , \quad D = D_0 + kD_1 + \cdots .$$

At zero order in k, the condition (3.3.3) is satisfied identically, (3.3.6) gives $D_0 = 0$ and then, from (3.3.5), ϑ is constant. Hence, from (3.3.4), we may take $\delta z_0 = z_{0\eta}$. At first order, the condition (3.3.3) is again satisfied identically. It is straightforward (see [42]) to find D_1 from (3.3.6) and $\vartheta_{1\eta}$ from (3.3.5); use of these expressions in (3.3.3) at second order leads to the desired nonlinear dispersion relation for the perturbations in the form

$$r_0 + r_1\omega_1 + r_2\omega_1^2 = 0. \tag{3.3.7}$$

The coefficients r_0, r_1 and r_2 depend on z_1, z_2 and z_3 as defined in (3.2.4). It turns out that the dispersion relation (3.3.7) has real roots for ω_1 for both the families of solutions (corresponding to $c = 1$ and $c = -1$, respectively) derived in Section 3.2. Consequently, it is predicted that both families of solutions are stable to long wavelength perturbations. For the loop-like solutions, the existence of singular points at which the derivatives tend to infinity casts some doubts on the validity of the method. However, in [42] it is argued that, as the method depends on the average behavior over a wavelength, the method is indeed valid.

3.3.2 Ambiguous solutions

The ambiguous structure of the loop-like solutions is similar to the loop soliton solution to an equation that models a stretched rope [51]. Loop-like solitons on a vortex filament were investigated by Hasimoto [52] and Lamb, Jr [53]. From the mathematical point of view an ambiguous solution does not present difficulties whereas the physical interpretation of ambiguity always presents some difficulties. In this connection the problem of ambiguous solutions is regarded as important. The problem consists of whether the ambiguity has a physical nature or is related to the incompleteness of the mathematical model, in particular to the lack of dissipation.

We will consider the problem related to the singular points when dissipation takes place. At these points the dissipative term $\alpha\frac{\partial u}{\partial x}$ tends to infinity. The question arises: are there solutions of the equation (3.1.6) in a loop-like form? That the dissipation is likely to destroy the loop-like solutions can be associated with the following well-known fact [27]. For the simplest nonlinear equation without dispersion and dissipation, namely

$$\frac{\partial u}{\partial t} + u\frac{\partial u}{\partial x} = 0, \tag{3.3.8}$$

any initial smooth solution with boundary conditions

$$u|_{x\to+\infty} = 0, \qquad u|_{x\to-\infty} = u_0 = \text{const.} > 0$$

becomes ambiguous in the final analysis. When dissipation is considered, we have the Burgers equation [54]

$$\frac{\partial u}{\partial t} + u\frac{\partial u}{\partial x} + \mu\frac{\partial^2 u}{\partial x^2} = 0.$$

The dissipative term in this equation and Eq. (2.2.5) for low frequency are coincident. The inclusion of the dissipative term transforms the solutions so that they cannot be ambiguous as a result of evolution. The wave parameters are always unambiguous. What happens in our case for high frequency when the dissipative term has the form αu (see Eq. (3.1.6))? Will the inclusion of dissipation give rise to unambiguous solutions?

We derive (see [11]) that the dissipative term, with dissipation parameter less than some limit value α^*, does not destroy the loop-like solutions.

Let us consider Eq. (3.1.6) in variables (3.2.1) with $v > 0$ and $c = 1$

$$z_{\eta\tau} + (zz_\eta)_\eta + (z+1) + \alpha z_\eta = 0. \tag{3.3.9}$$

We investigated the solution behavior within the neighborhood of singular points $z = 0$ where $z_\eta \to \pm\infty$ and $z_\tau \ll z_\eta$. Therefore in the investigated equation (3.3.9) we neglect the value z in the brackets $(z+1)$, and also omitted the term z_τ, to obtain

$$(zz_\eta)_\eta + 1 + \alpha z_\eta = 0. \tag{3.3.10}$$

It is convenient to use the inverse function $\eta = \eta(z)$. Taking into account $z_\eta = 1/\eta_z$ and $z_{\eta\eta} = -\eta_{zz}/\eta_z^3$, equation (3.3.10) is rewritten as

$$-z\eta_{zz} + \eta_z^3 + \alpha\eta_z^2 + \eta_z = 0.$$

Introducing the definition $q \equiv \eta_z$, this equation can be integrated to obtain

$$\int \frac{dq}{q(q^2 + \alpha q + 1)} = \int \frac{dz}{z}.$$

Depending on the sign of the quantity $1 - \alpha^2/4$, the latter expression has two different forms. We have introduced the critical value α^* of the parameter α defined by

$$\alpha^* = 2. \tag{3.3.11}$$

For $\alpha < \alpha^*$ (i.e. $1 - \alpha^2/4 > 0$), we get

$$\ln\left[\frac{z^2}{q^2}(q^2+\alpha q+1)\right] = -\frac{2\alpha}{\sqrt{4-\alpha^2}}\ \tan^{-1}\frac{2q+\alpha}{\sqrt{4-\alpha^2}} + \ln c_1, \tag{3.3.12}$$

and for $\alpha > \alpha^*$ (i.e. $1 - \alpha^2/4 < 0$), we have

$$\ln\left[\frac{z^2}{q^2}(q^2+\alpha q+1)\right] = \frac{\alpha}{\sqrt{\alpha^2-4}}\ \ln\left|\frac{2q+\alpha+\sqrt{\alpha^2-4}}{2q+\alpha-\sqrt{\alpha^2-4}}\right| + \ln c_2. \tag{3.3.13}$$

We analyze the expression (3.3.12). First let us verify the special case $\alpha = 0$. We have

$$\frac{z^2}{q^2}(q^2 + 1) = c_1,$$

or

$$z^2 + \frac{1}{4}(z^2)_\eta^2 = c_1.$$

Hence in the vicinity of $z = 0$

$$\eta + \eta_0 = \pm\frac{1}{2}\int \frac{dz^2}{\sqrt{c_1 - z^2}} = \mp\sqrt{c_1 - z^2}.$$

We arrive at the result given in [41], namely that with the lack of dissipation $\alpha = 0$ the integral curves pass over an ellipse at $z = 0$.

Now we investigate the case $0 < \alpha < \alpha^*$. It is easy to show that the r.h.s. of (3.3.12) is always bounded for any value $q \equiv z_\eta^{-1}$. In the neighborhood of $z = 0$ the r.h.s. of relation (3.3.12) is close to the value

$$-\frac{2\alpha}{\sqrt{4-\alpha^2}} \tan^{-1}\frac{\alpha}{\sqrt{4-\alpha^2}} + \ln c_1 \equiv \ln c_3.$$

Consequently, we arrive at the equation

$$\frac{z^2}{q^2}(q^2 + \alpha q + 1) = c_3.$$

Even not integrating this equation, it is easy to show that at $z = 0$ we must have $q = 0$ since in general $c_3 \neq 0$. This means that at $z = 0$ the derivatives have the values

$$\eta_z = 0, \qquad z_\eta = \pm\infty.$$

At $z = 0$ the solution becomes ambiguous.

In the case $\alpha > \alpha^*$ there is the solution

$$z = 0, \qquad q = \eta_z \neq 0, \qquad z_\eta \neq \pm\infty.$$

In fact, at $z = 0$ we obtain from the r.h.s. of (3.3.13)

$$q = \eta_z = -\tfrac{1}{2}(\alpha + \sqrt{\alpha^2 - 4}) \neq 0. \tag{3.3.14}$$

Thus, the derivative z_η at $z = 0$ is bounded by a finite value. The solution is always unambiguous.

Let us consider the solution behavior in the neighborhood of $z = 0$ as $\alpha \to \alpha^*$. We first consider the case $\alpha \to \alpha^* - 0$. According to (3.3.12) the r.h.s. of this equation tends to minus infinity, i.e. at $z \approx 0$ we have $q = \eta_z \neq 0$. Consequently, there is no loop-like solution.

When $\alpha \to \alpha^* + 0$ there is also a solution with $q = \eta_z \neq 0$ at $z = 0$. The root $q = 0$ at $z = 0$ seems possible in this case since (3.3.13) transforms to

$$\ln\left[\frac{z^2}{q^2}(q^2 + \alpha q + 1)\right] = \frac{2\alpha}{2q + \alpha} + \ln c_2, \tag{3.3.15}$$

However, as appears from (3.3.14), the r.h.s. of the equation (3.3.15) tends to minus infinity so that $q \neq 0$ at $z = 0$. Therefore, in the case $\alpha \to \alpha^*$ the dissipation destroys the loop-like solutions.

We have proved the following statement. For values of $\alpha < \alpha^*$ the inclusion of the dissipative term does not change the loop-like solutions of equation (3.1.6), while for $\alpha \geq \alpha^*$ there is no solution with an infinite gradient.

The common form of the dissipative term for high-frequency perturbations αu (which does not depend on the nature of the exchange processes) cannot preclude the possibility of a formation of a multi-valued solution from an initial single-valued profile. In this case there are infinite gradients in contrast to the profiles of a wave for the low frequencies when the dissipative term has the form $\beta \frac{\partial^2 u}{\partial x^2}$.

3.3.3 Interpretation of the ambiguous solutions

Now we give a physical interpretation to ambiguous solutions. Since the solution to the VE has a parametric form (3.2.6), (3.2.7) or (3.2.10), (3.2.11), there is a space of variables in which the solution is a single-valued function. Hence, we can solve the problem of the ambiguous solution. Several states with their thermodynamic parameters can occupy one microvolume. It is assumed that the interaction between the separated states occupying one microvolume can be neglected in comparison with the interaction between the particles of one thermodynamic state. Even if we take into account the interaction between the separated states in accordance with the dynamic state equation (2.1.6) then, for high frequencies, a dissipative term arises which is similar to the corresponding term in Eq. (2.2.6), but with the other relaxation time. In this sense the separated terms are distributed in space, but describing the wave process we consider them as interpenetratable. A similar situation, when several components with different hydrodynamic parameters occupy one microvolume, has been assumed in mixture theory (see, for instance [55, 56]). Such a fundamental assumption in the theory of mixtures is physically impossible (see [55], p.7), but it is appropriate in the sense that separated components are multi-velocity interpenetratable continua [57].

The KdV and KdVB equations are employed to describe a number of evolution processes when the low-frequency approach turns out to be adequate. In these cases thermodynamic parameters of a medium are close to the equilibrium values, the microvolume state is defined by one set of thermodynamic values, and the disturbance from the equilibrium is taken into account by means of an expansion in gradients [58]. If the low-order expansions within the framework of such an approach give rise to an inadequate description, we could take into account the terms of higher order and as a result consider higher frequencies. For example, if Eq. (3.3.8) has an ambiguous solution (or discontinuous solution), the improvement of models through adding higher degree derivatives excludes the ambiguous solutions. So, in the low-frequency approach, an ambiguity is connected with the incompleteness of the mathematical model.

In contrast to this, in models for the propagation of high-frequency perturbations, the disturbance from the frozen state is taken into account by means of an expansion in terms of an integral (see Eq. (3.1.4) and Eq. (3.1.6)). The integral terms contain the prehistory of the process. We have just established that a higher order of expansion (in particular, the dissipative term) for the high-frequency evolution equation still allows ambiguous solutions. Consequently, the ambiguity of solution does not relate to the incompleteness of the mathematical model, in particular to the lack of dissipation. In addition there is the space of new independent variables where the solution is a single-valued function.

Consequently, the following three observations show that, in the framework of the approach considered here, there are multi-valued solutions when we model high-frequency wave processes: (1) All parts of the loop-like solution are stable to perturbations. (2) Dissipation does not destroy the loop-like solutions. (3) The investigation regarding the interaction of the solitons has shown that it is necessary to take into account the whole ambiguous solution, and not just the separate parts.

3.4 Appendix

As mentioned in Section 3.2, provided that $0 \le A \le 1$, the real roots z_1, z_2 and z_3 (with $z_1 \le z_2 \le z_3$) of $f(z)$ in (3.2.4) are the roots of the cubic equation

$$z^3 + \tfrac{3}{2}cz^2 - \tfrac{1}{2}c^3 A = 0. \tag{A.1}$$

There are several ways of calculating these roots. We note that, as there is only one independent constant A in (A.1), only one constant needs to be specified in order to determine the roots. Here we consider three possible choices of this constant.

First, note that the trigonometrical solution to a cubic equation given in Section 52 [108] may be used to show that

$$z_1 = -\frac{c}{2} + |c| \cos\left(\frac{\theta + 2\pi}{3}\right), \tag{A.2}$$

$$z_2 = -\frac{c}{2} + |c| \cos\left(\frac{\theta - 2\pi}{3}\right), \tag{A.3}$$

$$z_3 = -\frac{c}{2} + |c| \cos\left(\frac{\theta}{3}\right), \tag{A.4}$$

where

$$\cos\theta = \mathrm{sgn}(c)(2A - 1). \tag{A.5}$$

Hence, given A with $0 \leq A \leq 1$, we can find θ from (A.5) and then z_1, z_2 and z_3 from (A.2) – (A.4).

Alternatively, by using (A.2) – (A.4) in $m = (z_3 - z_2)/(z_3 - z_1)$ we find that

$$\theta = 3\tan^{-1}\left[\frac{\sqrt{3}(1-m)}{1+m}\right]. \tag{A.6}$$

Hence, given m with $0 \leq m \leq 1$, we can find θ from (A.6) and then z_1, z_2 and z_3 from (A.2) – (A.4).

Finally, by dividing the trinomial by $z - z_3$, we find that

$$z_{2,1} = \tfrac{1}{2}\left(-q \pm \sqrt{q^2 - 4z_3 q}\right), \qquad q: = \tfrac{3}{2}\, c + z_3. \tag{A.7}$$

Hence, given z_3 with $z_3 \in [0, 0.5]$ for $c = 1$ or $z_3 \in [1, 1.5]$ for $c = -1$, we can find z_1 and z_2 from (A.7).

Chapter 4
Some equations related to the VE

The solution procedure, which was suggested in [41, 42] and used for the Vakhnenko equation (2.3.2) (see also Section 3.2), can be successfully adopted to find implicit periodic and solitary travelling-wave solutions of the Degasperis–Procesi equation in [69] (see also Section 4.2), the Camassa–Holm equation [71] (see also Section 4.3), the transformed Hirota–Satsuma-type 'shallow water wave' equation [72] and special cases thereof, namely the generalised Vakhnenko equation and the modified generalised Vakhnenko equation (see also Section 4.5), the short-pulse equation [103] (see also Section 4.6) and other equations. An important feature of the method is that it delivers solutions in which both the dependent variable and the independent variable are given in terms of a parameter.

4.1 The peakon b-family equation

Mikhailov and Novikov developed a powerful extension of the symmetry classification method [59]. Applying this to the equation

$$u_t - u_{txx} + (b+1)uu_x = bu_xu_{xx} + uu_{xxx}, \tag{4.1.1}$$

they found that only the cases with parameter $b = 2,\ 3$ could possess infinitely many commuting symmetries, and so only these two cases are integrable.

In [61] the family of equations was dubbed the 'peakon b-family'. The family of equations (4.1.1) with $b > 1$ was discussed in [60]. Phase portraits were used to categorize travelling-wave solutions. As discussed in [62], the family of Eq. (4.1.1) contains only two integrable equations, namely the dispersionless Camassa–Holm equation (CHE) for which $b = 2$ [63]

$$u_t - u_{txx} + 3uu_x = 2u_x u_{xx} + uu_{xxx}, \tag{4.1.2}$$

and the Degasperis–Procesi equation (DPE) for which $b = 3$ [64]

$$u_t - u_{txx} + 4uu_x = 3u_x u_{xx} + uu_{xxx}, \tag{4.1.3}$$

or in equivalent form

$$(u_t + uu_x)_{xx} = u_t + 4uu_x. \tag{4.1.4}$$

Originally Eq. (4.1.2) was derived as an equation for shallow water waves [63]. Later Chen et al. showed that Eq. (4.1.2) can be applied successfully to describe turbulent flows [65]. Since Hone and Wang [73] revealed the connection of the DPE and the Vakhnenko equation (see Section (4.1.2)), Eq. (4.1.3) can be used to model wave perturbations in relaxing media. As proved by Lenells (and it is important from the physical point of view), the multivalued solutions of the CHE and the DPE can be the basis for the construction of one-valued solutions [66].

It has been known for some time that the dispersionless Camassa–Holm equation has a weak solution in the form of a single 'peakon' [63]

$$u(x,t) = v \exp(-|x - vt|), \tag{4.1.5}$$

where v is a constant, and an N-peakon solution [67] that is just a superposition of peakons, namely

$$u(x,t) = \sum_{j=1}^{N} p_j(t) \exp(-|x - q_j(t)|), \tag{4.1.6}$$

where the $p_j(t)$ and $q_j(t)$ satisfy a certain associated dynamical system. (A substantial list of references regarding the properties of the CHE may be found in [66].) More recently Degasperis, Holm and Hone [62] proved the integrability of the DPE, and showed that the equation also has single and N-peakon solutions of the form (4.1.5) and (4.1.6) respectively; the peakon dynamics were discussed and compared with the analogous results for Camassa–Holm peakons.

A classification of travelling-wave solutions of the CHE was given in [66]. However, explicit solutions were given only for the solitary peakon and periodic peakon waves. Periodic smooth-hump waves and periodic cuspon waves were investigated numerically in [68].

4.1.1 An integrated form of Eq. (4.1.1)

In order to seek travelling-wave solutions to (4.1.1), it is convenient to introduce a new dependent variable z defined by

$$z = (u - v)/|v| \tag{4.1.7}$$

and to assume that z is an implicit or explicit function of η, where $\eta = x - vt - x_0$, v and x_0 are arbitrary constants, and $v \neq 0$. Then (4.1.1) becomes

$$zz_{\eta\eta\eta} + bz_\eta z_{\eta\eta} - (b+1)zz_\eta - bcz_\eta = 0, \quad \text{with} \quad c = v/|v| = \pm 1. \tag{4.1.8}$$

After two integrations (4.1.8) gives

$$(zz_\eta)^2 = f(z), \tag{4.1.9}$$

where

$$f(z) = z^4 + 2cz^3 + Az^2 + Bz^{3-b}, \tag{4.1.10}$$

and A and B are real constants.

Note that for $b > 1$, $f(z)$ is a quartic for $b = 2$ or $b = 3$ only. This explains why the technique that was used for the DPE in [69] (see also Section 4.2) also works for the CHE [71] (see also Section 4.3).

For the case $b = 3$, when (4.1.1) is the DPE (4.1.3), and for the case $b = 2$, when (4.1.1) is the CHE (4.1.2) we will consider the travelling-wave solutions in Section (4.2) and Section (4.3), respectively.

4.1.2 Connection of the VE with the DPE

Hone and Wang [73] have shown that there is a subtle connection between the Sawada–Kotera hierarchy and the VE, between the DPE and the VE, and between the Lax pairs of the DPE and VE. In particular they noted that the application of the transformations

$$x \to \tilde{\varepsilon}x - \frac{t}{3\tilde{\varepsilon}}, \qquad t \to \tilde{\varepsilon}t, \qquad u \to u - \frac{1}{3\tilde{\varepsilon}^2} \tag{4.1.11}$$

to the DPE (4.1.3), where $\tilde{\varepsilon}$ is a real positive constant, results in

$$((u_t + uu_x)_x + u)_x = \tilde{\varepsilon}^2(u_t + 4uu_x). \tag{4.1.12}$$

In the limit $\tilde{\varepsilon} \to 0$, (4.1.12) reduces to the derivative of the VE (2.3.2)

$$(u_t + uu_x)_x + u = 0.$$

We will refer to (4.1.12) as the transformed DPE (see also Section 4.2.2).

The VE has a loop-like solution, hence the DPE should admit a loop-like solution. However, it engages our attention that soliton solutions have not been observed for the DPE recently.

4.2 The Degasperis–Procesi equation

Now we investigate the travelling-wave solutions of the DPE and the transformed DPE. We show that the solutions are characterized by two parameters. Hump-like, loop-like and coshoidal periodic-wave solutions are found; hump-like, loop-like and peakon solitary-wave solutions are obtained as well [69, 70]. In an appropriate limit the solutions of the DPE lead to the solutions of the VE.

In Section 4.2.1 we show that, for travelling-wave solutions, the DPE may be reduced to a first order ODE involving two arbitrary constants A and B. We show that there are four distinct periodic solutions corresponding to four different ranges of values of A; for a given allowed value of A, B is restricted to a range of values. By using results established in the Appendix to this Chapter (see Section 4.7), we express the periodic solutions in implicit form; these solutions involve elliptic integrals and Jacobian elliptic functions with parameter m, where $0 < m < 1$. We also investigate the limiting form of these solutions when $m = 1$.

In Section 4.2.2 we perform the corresponding analysis for the transformed DPE. We consider the case for which the first-order ODE to which the transformed DPE may be reduced involves only a single integration constant B. We find that there are eight distinct solution regimes corresponding to four different ranges of values of $\tilde{\varepsilon}^2$ and to the two possible directions of propagation. In each case B is restricted to a range of values. We show that, when $\tilde{\varepsilon} \to 0$ in Eq. (4.1.12), the periodic and solitary-wave solutions to the VE are recovered.

4.2.1 Solutions of the DPE

In this Section we seek travelling-wave solutions of the DPE (4.1.3). Note that there are no bound stationary solutions of (4.1.3) that are in the form $u = u(x)$. That being the case, it is convenient to introduce the new variables z and η as defined in Section 4.1.1. Then (4.1.3) becomes

$$(zz_\eta)_{\eta\eta} = (4z+3c)z_\eta, \text{ where } c: = \frac{v}{|v|} = \pm 1. \tag{4.2.1}$$

After two integrations, (4.2.1) is reduced to

$$(zz_\eta)^2 = f(z), \tag{4.2.2}$$

where $f(z)$ is the polynomial given by (4.1.10) with $b = 3$. This polynomial can be written in terms of the roots of the equation $f(z) = 0$ as follows:

$$f(z): = z^4 + 2cz^3 + Az^2 + B \equiv (z-z_1)(z-z_2)(z_3-z)(z_4-z), \tag{4.2.3}$$

where A and B are real constants. For the solutions that we are seeking, z_1, z_2, z_3 and z_4 are real constants with $z_1 \le z_2 \le z \le z_3 \le z_4$ and $z_1 + z_2 + z_3 + z_4 = -2c$. Equation (4.2.2) is of the same form as (A.1) in the Appendix to this Chapter (see Section 4.7) with $\varepsilon = 1$. Hence we can make use of the solutions given in the Appendix, but with $\varepsilon = 1$.

Note that (4.2.1) is invariant under the transformation $z \to -z$, $c \to -c$; this corresponds to the transformation $u \to -u$, $v \to -v$. Here we will seek the family of solutions of (4.2.1) for which $v > 0$ and so, from here on in this Section, we set $c = 1$.

For convenience we define $g(z)$ and $h(z)$ by

$$f(z) = z^2 g(z) + B, \quad \text{where} \quad g(z): = z^2 + 2z + A, \tag{4.2.4}$$

and

$$f'(z) = 2zh(z), \quad \text{where} \quad h(z): = 2z^2 + 3z + A, \tag{4.2.5}$$

and define z_L, z_U, B_L and B_U by

$$z_L: = -\frac{1}{4}(3 + \sqrt{9-8A}\,), \quad z_U: = -\frac{1}{4}(3 - \sqrt{9-8A}\,), \tag{4.2.6}$$

$$B_L: = -z_L^2 g(z_L) = -\frac{A^2}{4} + \quad \frac{9A}{8} \quad \frac{27}{32} + \frac{1}{32}(9-8A)\sqrt{9-8A}\,, \tag{4.2.7}$$

$$B_U: = -z_U^2 g(z_U) = -\frac{A^2}{4} + \quad \frac{9A}{8} \quad \frac{27}{32} - \frac{1}{32}(9-8A)\sqrt{9-8A}\,; \tag{4.2.8}$$

z_L and z_U are the roots of $h(z) = 0$.

Provided A is non-zero and is such that $A < \frac{9}{8}$, $f(z)$ has three distinct stationary points that occur at $z = z_L$, $z = z_U$ and $z = 0$, and comprise two minimums separated by a maximum. Four cases are possible for the polynomial $f(z)$ corresponding to different ranges of values of A (see Fig. 4.1). In this case (4.2.2) has periodic and solitary-wave solutions that have different analytical forms depending on the values of A and B as described in Sections 4.2.1 a–d as follows.

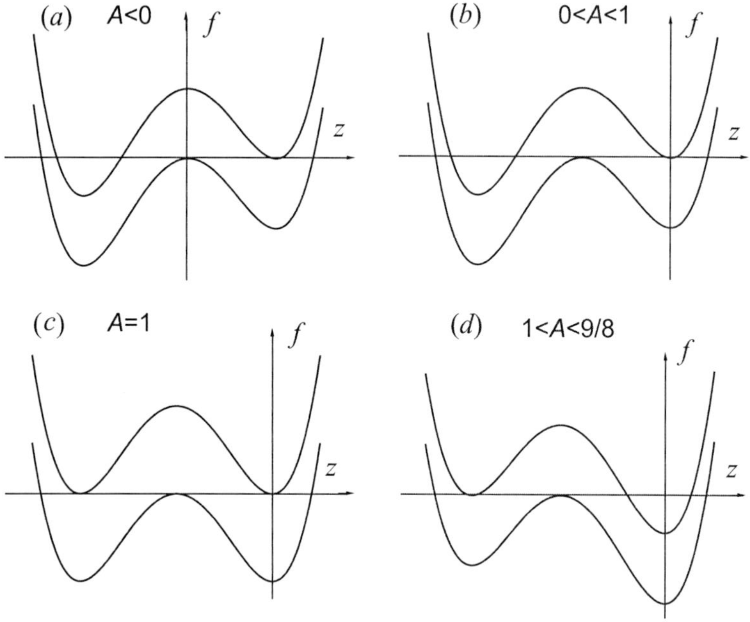

Figure 4.1: Four possible cases for the polynomial $f(z)$.

4.2.1 a. $A < 0$

In this case $z_L < 0 < z_U$ with $f(z_L) < f(z_U)$; see Fig. 4.1(a), where the lower and upper curves correspond to $B = 0$ and $B = B_U$, respectively. For each value of A satisfying $A < 0$ there are periodic inverted loop solutions to (4.2.2) given by (A.5) and (A.7) with $0 < B < B_U$ so that $0 < m < 1$, and with wavelength given by (A.8); see Fig. 4.2(a) for an example.

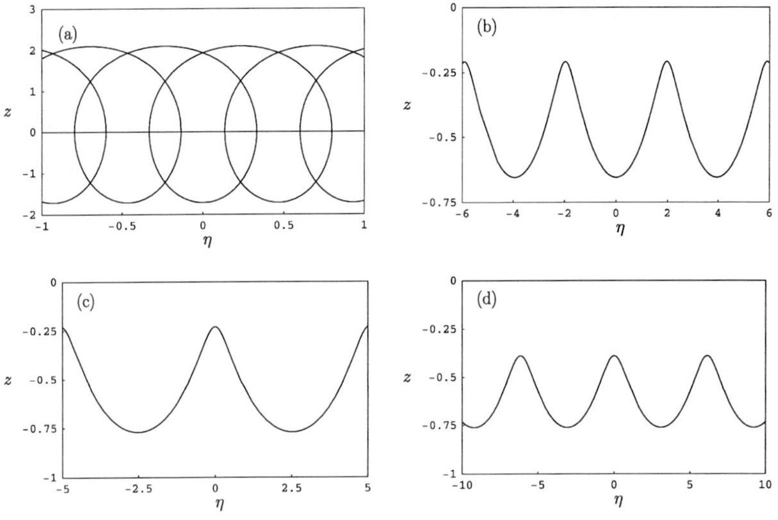

Figure 4.2: Periodic solutions of the DPE with $0 < m < 1$: (a) $A = -27$, $B = 0.75B_U$ so $m = 0.842$, $\lambda = 0.466$; (b) $A = 15/16$, $B = 0.5B_U$ so $m = 0.746$, $\lambda = 3.941$; (c) $A = 1$, $B = 0.5B_U = -1/32$ so $m = 0.746$, $\lambda = 5.038$; (d) $A = 135/128$, $B = 0.5(B_U + B_L)$ so $m = 0.729$, $\lambda = 6.140$.

$B = B_U$ corresponds to the limit $z_3 = z_4 = z_U$ so that $m = 1$, and then the solution is an inverted loop-like solitary wave given by (A.9) with $z_2 \leq z < z_U$ and

$$z_1 = -\frac{1}{4}\left(1 + \sqrt{9 - 8A}\right) - \frac{1}{2}\sqrt{1 + \sqrt{9 - 8A}}, \tag{4.2.9}$$

$$z_2 = -\frac{1}{4}\left(1 + \sqrt{9 - 8A}\right) + \frac{1}{2}\sqrt{1 + \sqrt{9 - 8A}}; \tag{4.2.10}$$

see Fig. 4.3(a) for an example. Note that $z_2 \to 0$ and $z_U \to 0$ as $A \to 0$. The amplitude $z_U - z_2$ of the solitary wave increases from 0 as $|A|$ increases from 0.

The loop-like nature of the solitary wave is due to the fact that $z = 0$ is in the range $z_2 \leq z < z_U$. For small z, (4.2.2) gives $z_\eta \simeq \pm\sqrt{B_U}/z$ and so $|z_\eta| \to \infty$ as $z \to 0$. It follows that the solution curve (see Fig. 4.3(a) for example) is normal to the η axis at the points $(\mp W/2, 0)$, where W is the maximum width of the loop; from

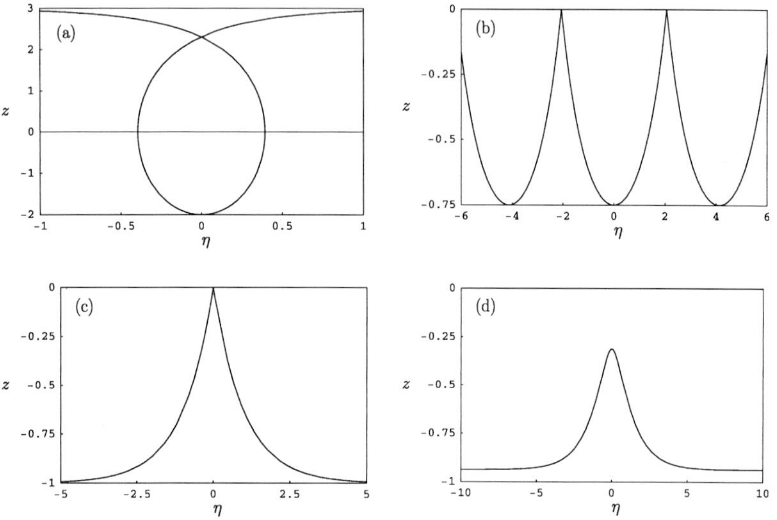

Figure 4.3: Solutions of the DPE with $m = 1$: (a) $A = -27$, $B = B_U$, $W = 0.788$; (b) $A = 15/16$, $B = 0$, $\lambda = 4.127$; (c) $A = 1$, $B = 0$; (d) $A = 135/128$, $B = B_L$.

(A.9), W is given by

$$W = 4 \tanh^{-1}\left(\sqrt{\frac{z_2}{z_1}}\right) - \frac{2z_U}{p} \tanh^{-1}\left(\sqrt{\frac{z_2}{nz_1}}\right). \qquad (4.2.11)$$

W increases from 0 as $|A|$ increases from 0. Near the points $(\mp W/2, 0)$ the loop is approximately parabolic and given by

$$z^2 \simeq 2\sqrt{B_U}\left(\frac{W}{2} \pm \eta\right). \qquad (4.2.12)$$

4.2.1 b. $0 < A < 1$

In this case $z_L < z_U < 0$ with $f(z_L) < f(0)$; see Fig. 4.1(b), where the lower and upper curves correspond to $B = B_U$ and $B = 0$, respectively. For each value of A satisfying $0 < A < 1$ there are periodic hump solutions to (4.2.2) given by (A.5) and (A.7) with $B_U < B < 0$ so that $0 < m < 1$, and with wavelength given by (A.8); see Fig. 4.2(b) for an example.

$B = 0$ corresponds to the limit $z_3 = z_4 = 0$ so that $m = 1$, and then the solution has $z_2 \le z \le 0$ and is given by (A.9) with z_1 and z_2 given by the roots of $g(z) = 0$, namely

$$z_1 = -1 - \sqrt{1-A}\,, \quad z_2 = -1 + \sqrt{1-A}\,. \tag{4.2.13}$$

In this case we obtain a weak solution, namely the periodic upward spike

$$z = z(\eta - 2j\eta_m), \quad (2j-1)\eta_m \le \eta \le (2j+1)\eta_m, \quad j = 0, \pm 1, \pm 2, \ldots, \tag{4.2.14}$$

where

$$z(\eta) := [z_2 - z_1 \tanh^2(\eta/2)] \cosh^2(\eta/2) \equiv -1 + \sqrt{1-A}\, \cosh\eta \tag{4.2.15}$$

and

$$\eta_m = 2\tanh^{-1}\left(\sqrt{\frac{z_2}{z_1}}\right) \equiv \cosh^{-1}\left(\frac{1}{\sqrt{1-A}}\right); \tag{4.2.16}$$

see Fig. 4.3(b) for an example. (4.2.16) is similar in form to the spatially periodic solution of the Camassa–Holm equation that has been dubbed a 'coshoidal wave' by Boyd [109]. Note that $z_2 \to 0$ and $\eta_m \to 0$ as $A \to 0$, and that $z_2 \to -1$ and $\eta_m \to \infty$ as $A \to 1$. Hence the amplitude $|z_2|$ of the coshoidal wave (4.2.16) increases from 0 to 1 as A increases from 0 to 1, and its wavelength $\lambda := 2\eta_m$ increases from 0 to infinity.

4.2.1 c. $A = 1$

In this case $z_L < z_U < 0$ with $f(z_L) = f(0)$; see Fig. 4.1(c), where the lower and upper curves correspond to $B = B_U$ and $B = 0$, respectively. For $A = 1$ there are periodic hump solutions to (4.2.2) given by (A.11) and (A.12) with $B_U < B < 0$ so that $0 < m < 1$, where $B_U = -1/16$, and with wavelength given by (A.13); see Fig. 4.2(c) for an example. An alternative solution is given by (A.5) and (A.7); this is just the former solution phase-shifted by $\lambda/2$.

$B = 0$ corresponds to the limit $z_1 = z_2 = z_L = -1$ and $z_3 = z_4 = 0$. In this case neither (A.9) nor (A.14) is appropriate. Instead we consider (4.2.2) with $f(z) = z^2(z+1)^2$ and note that the bound solution has $-1 < z \leq 0$. On integrating (4.2.2) and setting $z = 0$ at $\eta = 0$ we obtain the weak solution

$$z = e^{-|\eta|} - 1, \tag{4.2.17}$$

i.e. a single peakon with amplitude 1; see Fig. 4.3(c). In terms of the original dependent variable u, (4.2.17) is equivalent to (4.1.5) with $v > 0$.

4.2.1 d. $1 < A < 9/8$

In this case $z_L < z_U < 0$ with $f(z_L) > f(0)$; see Fig. 4.1(d), where the lower and upper curves correspond to $B = B_U$ and $B = B_L$, respectively. For each value of A satisfying $1 < A < \frac{9}{8}$ there are periodic hump solutions to (4.2.2) given by (A.11) and (A.12) with $B_U < B < B_L$ so that $0 < m < 1$, and with wavelength given by (A.13); see Fig. 4.2(d) for an example.

$B = B_L$ corresponds to the limit $z_1 = z_2 = z_L$ so that $m = 1$, and then the solution is a hump-like solitary wave given by (A.14) with $z_L < z \leq z_3$ and

$$z_3 = -\frac{1}{4}\left(1 - \sqrt{9 - 8A}\right) - \frac{1}{2}\sqrt{1 - \sqrt{9 - 8A}}, \tag{4.2.18}$$

$$z_4 = -\frac{1}{4}\left(1 - \sqrt{9 - 8A}\right) + \frac{1}{2}\sqrt{1 - \sqrt{9 - 8A}}; \tag{4.2.19}$$

see Fig. 4.3(d) for an example. Note that $z_L \to -1$ and $z_3 \to 0$ as $A \to 1$, and that $z_L \to -3/4$ and $z_3 \to -3/4$ as $A \to 9/8$. The amplitude $z_3 - z_L$ of the solitary wave decreases from 1 to 0 as A increases from 1 to 9/8.

4.2.2 Solutions of the transformed DPE

In this Section we seek travelling-wave solutions of the transformed DPE (4.1.12). Note that there are no bound stationary solutions of (4.1.12) that are in the form $u = u(x)$. That being the case, it is convenient to introduce a new dependent variable z defined by

$$z = (u - v)/|v| \tag{4.2.20}$$

and to assume that z is an implicit or explicit function of η, where

$$\eta = (x - vt - x_0)/|v|^{1/2}, \tag{4.2.21}$$

v and x_0 are arbitrary constants, and $v \neq 0$. Then, with $\varepsilon = \tilde{\varepsilon}|v|^{1/2}$, (4.1.12) becomes

$$\Big((zz_\eta)_\eta + z + c\Big)_\eta = \varepsilon^2(4z + 3c)z_\eta, \qquad \text{where } c := \frac{v}{|v|} = \pm 1. \quad (4.2.22)$$

After one integration (4.2.22) gives

$$(zz_\eta)_\eta + z + c = \varepsilon^2(2z^2 + 3cz) + c_1\,, \tag{4.2.23}$$

where c_1 is an arbitrary real constant. Note that, in terms of z and η, the VE (2.3.2) becomes (3.2.3), namely

$$(zz_\eta)_\eta + z + c = 0. \tag{4.2.24}$$

The solitary-wave solution to (4.2.24) is such that $z_\eta \to 0$, $z_{\eta\eta} \to 0$ and $z + c \to 0$, as $|\eta| \to \infty$. We choose c_1 in (4.2.23) so that these conditions are satisfied. Accordingly, here we restrict attention to the particular case in which $c_1 = c^2\varepsilon^2$. Then, after one integration, (4.2.23) gives

$$(zz_\eta)^2 = \varepsilon^2 f(z), \tag{4.2.25}$$

where

$$f(z) := z^4 - \frac{2}{3\varepsilon^2}(1 - 3c\varepsilon^2)z^3 - \frac{c}{\varepsilon^2}(1 - c\varepsilon^2)z^2 + B \equiv \tag{4.2.26}$$

$$\equiv (z - z_1)(z - z_2)(z_3 - z)(z_4 - z)$$

and B is a real constant. For the solutions that we are seeking, z_1, z_2, z_3 and z_4 are real constants with $z_1 \leq z_2 \leq z \leq z_3 \leq z_4$. Equation (4.2.25) is of the same form as (A.1) in the Appendix to this Chapter (see Section 4.7). Hence we can make use of the solutions given in the Appendix.

For convenience we define $g(z)$ and $h(z)$ by

$$f(z) = z^2 g(z) + B, \quad \text{where} \quad g(z) := z^2 - \frac{2}{3\varepsilon^2}(1 - 3c\varepsilon^2)z - \frac{c}{\varepsilon^2}(1 - c\varepsilon^2),$$

and (4.2.27)

$$f'(z) = 2zh(z), \quad \text{where} \quad h(z) := 2z^2 - \frac{1}{\varepsilon^2}(1 - 3c\varepsilon^2)z - \frac{c}{\varepsilon^2}(1 - c\varepsilon^2),$$

(4.2.28)

and define z_L, z_U, B_L and B_U by

$$z_L\colon = -c, \qquad z_U\colon = \frac{1}{2\varepsilon^2}\left(1 - c\varepsilon^2\right), \tag{4.2.29}$$

$$B_L\colon = -z_L^2 g(z_L) = \frac{c}{3\varepsilon^2}, \tag{4.2.30}$$

$$B_U\colon = -z_U^2 g(z_U) = \frac{1}{48\varepsilon^8}(1 + 3c\varepsilon^2)(1 - c\varepsilon^2)^3; \tag{4.2.31}$$

z_L and z_U are the roots of $h(z) = 0$.

Provided that $\varepsilon^2 > 0$ and $\varepsilon^2 \neq 1$, $f(z)$ has three distinct stationary points that occur at $z = z_L$, $z = z_U$ and $z = 0$, and comprise two minimums separated by a maximum. In this case (4.2.25) has periodic and solitary-wave solutions that have different analytical forms depending on the values of ε^2 and B as described in Sections 4.4.2 a–h as follows.

4.2.2 a. $c = 1$, $\varepsilon^2 > 1$

In this case $z_L < z_U < 0$ with $f(z_L) < f(0)$. For each value of ε^2 satisfying $\varepsilon^2 > 1$ there are periodic hump solutions to (4.2.25) given by (A.5) and (A.7) with $B_U < B < 0$ so that $0 < m < 1$, and with wavelength given by (A.8); see Fig. 4.4(a) for an example.

$B = 0$ corresponds to the limit $z_3 = z_4 = 0$ so that $m = 1$, and then the solution has $z_2 \leq z \leq 0$ and is given by (A.9) with z_1 and z_2 given by the roots of $g(z) = 0$, where g is defined in (4.2.27), namely

$$z_1 = \frac{1}{3\varepsilon^2}\left(1 - 3\varepsilon^2 - \sqrt{1 + 3\varepsilon^2}\right), \quad z_2 = \frac{1}{3\varepsilon^2}\left(1 - 3\varepsilon^2 + \sqrt{1 + 3\varepsilon^2}\right). \tag{4.2.32}$$

In this case we obtain a weak solution, namely the coshoidal wave

$$z = z(\eta - 2j\eta_m), \quad (2j-1)\eta_m \leq \eta \leq (2j+1)\eta_m, \quad j = 0, \pm 1, \pm 2, \ldots, \tag{4.2.33}$$

where

$$z(\eta)\colon = [z_2 - z_1 \tanh^2(\varepsilon\eta/2)] \cosh^2(\varepsilon\eta/2) \equiv \tag{4.2.34}$$

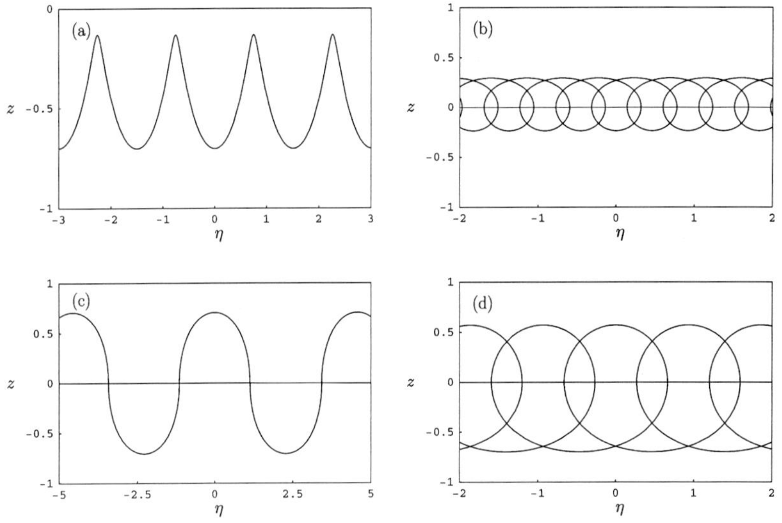

Figure 4.4: Periodic solutions of the transformed DPE with $c = 1$ and $0 < m < 1$: (a) $\varepsilon^2 = 8$, $B = 0.25B_U$ so $m = 0.869$, $\lambda = 1.507$; (b) $\varepsilon^2 = 1/2$, $B = 0.6B_U$ so $m = 0.730$, $\lambda = 0.458$; (c) $\varepsilon^2 = 1/3$, $B = 0.75$ so $m = 0.928$, $\lambda = 4.562$; (d) $\varepsilon^2 = 1/4$, $B = 0.75B_L$ so $m = 0.842$, $\lambda = 0.932$.

$$\equiv \frac{1}{3\varepsilon^2}\left(1 - 3\varepsilon^2 + \sqrt{1+3\varepsilon^2}\,\cosh(\varepsilon\eta)\right)$$

and

$$\eta_m = \frac{2}{\varepsilon}\tanh^{-1}\left(\sqrt{\frac{z_2}{z_1}}\right) \equiv \frac{1}{\varepsilon}\cosh^{-1}\left(\frac{3\varepsilon^2 - 1}{\sqrt{1+3\varepsilon^2}}\right); \qquad (4.2.35)$$

see Fig. 4.5(a) for an example. Note that $z_2 \to 0$ as $\varepsilon^2 \to 1$, and that $z_2 \to -1$ as $\varepsilon^2 \to \infty$. Hence the amplitude $|z_2|$ of the coshoidal wave increases from 0 to 1 as ε^2 increases from 1 to infinity. As ε^2 increases from 1, the wavelength $\lambda := 2\eta_m$ increases from 0, reaches a maximum value of 1.827 at $\varepsilon^2 = 2.769$, and then decreases to 0 as $\varepsilon^2 \to \infty$.

4.2.2 b. $c = 1,\ 1/3 < \varepsilon^2 < 1$

In this case $z_L < 0 < z_U$ with $f(z_L) < f(z_U)$. For each value of ε^2 satisfying $\frac{1}{3} < \varepsilon^2 < 1$ there are periodic inverted loop solutions to

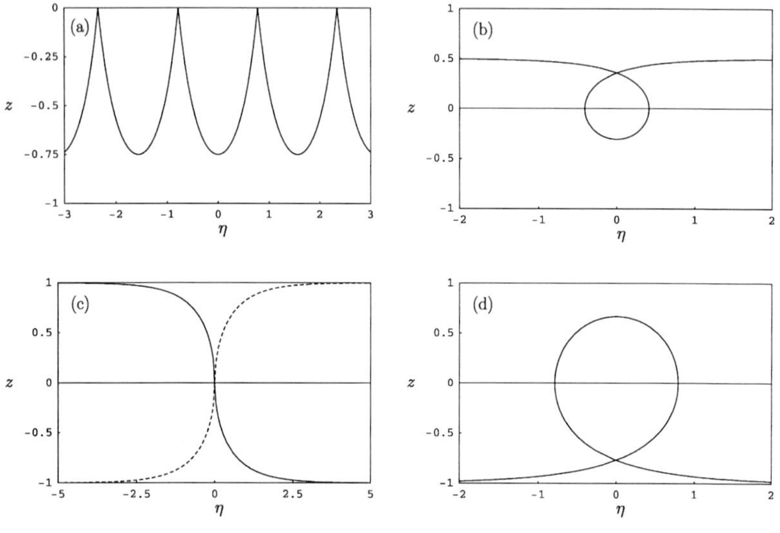

Figure 4.5: Solutions of the transformed DPE with $c = 1$ and $m = 1$: (a) $\varepsilon^2 = 8$, $B = 0$, $\lambda = 1.561$; (b) $\varepsilon^2 = 1/2$, $B = B_U$, $W = 0.818$; (c) $\varepsilon^2 = 1/3$, $B = 1$; (d) $\varepsilon^2 = 1/4$, $B = B_L$, $W = 1.577$.

(4.2.25) given by (A.5) and (A.7) with $0 < B < B_U$ so that $0 < m < 1$, and with wavelength given by (A.8); see Fig. 4.4(b) for an example.

$B = B_U$ corresponds to the limit $z_3 = z_4 = z_U$ so that $m = 1$, and then the solution is an inverted loop-like solitary wave given by (A.9) with $z_2 \le z < z_U$ and

$$z_1 = \frac{1}{6\varepsilon^2}\left(-1 - 3\varepsilon^2 - \sqrt{2(9\varepsilon^4 - 1)}\right), \tag{4.2.36}$$

$$z_2 = \frac{1}{6\varepsilon^2}\left(-1 - 3\varepsilon^2 + \sqrt{2(9\varepsilon^4 - 1)}\right);$$

see Fig. 4.5(b) for an example. The maximum width W of the loop is

$$W = \frac{1}{\varepsilon}\left[4\tanh^{-1}\left(\sqrt{\frac{z_2}{z_1}}\right) - \frac{2z_U}{p}\tanh^{-1}\left(\sqrt{\frac{z_2}{nz_1}}\right)\right]. \tag{4.2.37}$$

Note that $z_2 \to 0$ and $z_U \to 0$ as $\varepsilon^2 \to 1$, and that $z_2 \to -1$ and $z_U \to 1$ as $\varepsilon^2 \to \frac{1}{3}$. As ε^2 decreases from 1 to $\frac{1}{3}$, the amplitude $z_U - z_2$ of the solitary wave increases from 0 to 2, and W increases from 0 to infinity.

4.2.2 c. $c = 1$, $\varepsilon^2 = 1/3$

In this case $z_L < 0 < z_U$ with $f(z_L) = f(z_U)$. The z^3 term in the expression for $f(z)$ given by (4.2.26) is not present and hence $f(z)$ is even so that, for $0 < B < 1$ (with $B_U = B_L = 1$), $z_1 = -z_4$ and $z_2 = -z_3$. Then from the definition of m in (A.6) and the definitions of n in (A.5) or (A.11) we obtain the relation

$$m + n^2 - 2n = 0. \tag{4.2.38}$$

With (4.2.38), the results 141.01 and 414.01 in [44] may be used to show that λ given by (A.8) or (A.13) is zero, and hence that η given by (A.7) or (A.12) is periodic in w with period $2K$, where $K\!: = K(m)$ and $K(m)$ is the complete elliptic integral of the first k ind. It follows that, for each value of B such that $0 < B < 1$, the solution to (4.2.25) given by (A.5) and (A.7), or by (A.11) and (A.12), is just a closed curve around the origin in the z-η plane. This curve is symmetrical with respect to z and η and has infinite slope at the two points where $z = 0$. A periodic bell solution to (4.2.25), with wavelength $\lambda\!: = 4\eta(3K/2)$, may be constructed in parametric form as follows:

$$z = z(w) \tag{4.2.39}$$

$$\eta = \begin{cases} \eta(w) + (2+4j)\eta(3K/2), -K/2 + 2jK \le w \le K/2 + 2jK \\ \eta(w) + 4j\eta(3K/2), \quad K/2 + 2jK \le w \le 3K/2 + 2jK \end{cases} \tag{4.2.40}$$

where $z(w)$ and $\eta(w)$ are given by (A.5) and (A.7) respectively, and $j = 0, \pm 1, \pm 2, \ldots$; see Fig. 4.4(c) for an example.

$B = B_L = B_U = 1$ corresponds to the limit $z_1 = z_2 = z_L = -1$ and $z_3 = z_4 = z_U = 1$. In this case neither (A.9) nor (A.14) is appropriate. Instead we consider (4.2.25) with $f(z) = (z+1)^2(1-z)^2$ and note that the bound solutions have $-1 < z < 1$. On integrating (4.2.25) and setting $z = 0$ at $\eta = 0$ we find that there are two such solutions, namely the kink-like solitary waves

$$z = \begin{cases} +\sqrt{1 - \exp(-2|\eta|/\sqrt{3}\,)}, & \eta < 0, \\ -\sqrt{1 - \exp(-2|\eta|/\sqrt{3}\,)}, & \eta > 0, \end{cases} \tag{4.2.41}$$

and

$$z = \begin{cases} -\sqrt{1-\exp(-2|\eta|/\sqrt{3})}, & \eta < 0, \\ +\sqrt{1-\exp(-2|\eta|/\sqrt{3})}, & \eta > 0; \end{cases} \tag{4.2.42}$$

see Fig. 4.5(c) in which the solid and dashed curves correspond to (4.2.41) and (4.2.42) respectively.

4.2.2 d. $c = 1$, $0 < \varepsilon^2 < 1/3$

In this case $z_L < 0 < z_U$ with $f(z_U) < f(z_L)$. For each value of ε^2 satisfying $0 < \varepsilon^2 < \frac{1}{3}$ there are periodic loop solutions to (4.2.25) given by (A.11) and (A.12) with $0 < B < B_L$ so that $0 < m < 1$, and with wavelength given by (A.13); see Fig. 4.4(d) for an example. For a given choice of B, it is easy to verify numerically that, as ε^2 is made ever smaller (but finite), the aforementioned solution tends to the solution given by (3.2.10) and (3.2.11) with $\frac{1}{3}c^3 A = \varepsilon^2 B$ in (3.2.4); in other words the periodic loop solution of the VE for the case $v > 0$ is recovered in the limit $\varepsilon^2 \to 0$.

$B = B_L$ corresponds to the limit $z_1 = z_2 = z_L = -1$ so that $m = 1$, and then the solution is a loop-like solitary wave given by (A.14) with $-1 < z \leq z_3$ and

$$z_3 = \frac{1}{3\varepsilon^2}\left(1 - \sqrt{1-3\varepsilon^2}\right), \quad z_4 = \frac{1}{3\varepsilon^2}\left(1 + \sqrt{1-3\varepsilon^2}\right); \tag{4.2.43}$$

see Fig. 4.5(d) for an example. The maximum width W of the loop is

$$W = \frac{1}{\varepsilon}\left[4\tanh^{-1}\left(\sqrt{\frac{z_3}{z_4}}\right) - \frac{2}{p}\tanh^{-1}\left(\sqrt{\frac{z_3}{nz_4}}\right)\right]. \tag{4.2.44}$$

In the limit $\varepsilon^2 \to 0$, it is straightforward to show analytically that the solitary-wave solution reduces to (3.2.13) and that (4.2.44) reduces to (3.2.15); hence, as expected, the loop-like solitary-wave solution of the VE for the case $v > 0$ is recovered.

As ε^2 increases from 0 to $1/3$, the amplitude $z_3 + 1$ of the solitary wave increases from $3/2$ to 2, and W increases from the value given by (3.2.15), namely 0.8302, to infinity.

4.2.2 e. $c = -1$, $0 < \varepsilon^2 < 1/3$

In this case $0 < z_L < z_U$ with $f(0) > f(z_U)$. For each value of ε^2 satisfying $0 < \varepsilon^2 < 1/3/$ there are periodic well solutions to (4.2.25) given by (A.11) and (A.12) with $B_L < B < 0$ so that $0 < m < 1$, and with wavelength given by (A.13); see Fig. 4.6(a) for an example. For a given choice of B, it is easy to verify numerically that, as ε^2 is made ever smaller (but finite), the aforementioned solution tends to the solution given by (3.2.10) and (3.2.11) with $\frac{1}{3}c^3 A = \varepsilon^2 B$ in (3.2.4); in other words the periodic well solution of the VE for the case $v < 0$ is recovered in the limit $\varepsilon^2 \to 0$.

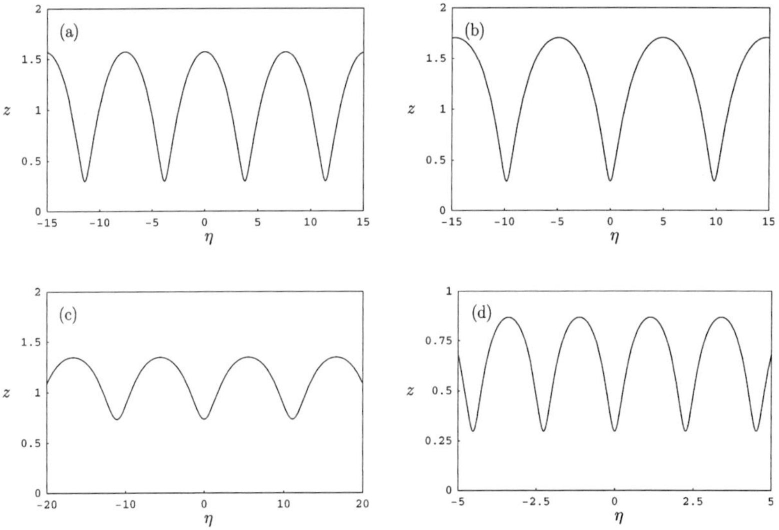

Figure 4.6: Periodic solutions of the transformed DPE with $c = -1$ and $0 < m < 1$: (a) $\varepsilon^2 = 1/4$, $B = 0.25B_L$ so $m = 0.842$, $\lambda = 7.600$; (b) $\varepsilon^2 = 1/3$, $B = 0.25B_L = -0.25$ so $m = 0.928$, $\lambda = 9.809$; (c) $\varepsilon^2 = 1/2$, $B = 0.25B_L + 0.75B_U$ so $m = 0.803$, $\lambda = 11.103$; (d) $\varepsilon^2 = 8$, $B = 0.25B_U + 0.75B_L$ so $m = 0.869$, $\lambda = 2.258$.

$B = 0$ corresponds to the limit $z_1 = z_2 = 0$ so that $m = 1$, and then the solution has $0 \le z \le z_3$ and is given by (A.14) with z_3 and z_4 given by the roots of $g(z) = 0$, where g is defined in (4.2.27), namely

$$z_3 = \frac{1}{3\varepsilon^2}\left(1 + 3\varepsilon^2 - \sqrt{1 - 3\varepsilon^2}\right), \quad z_4 = \frac{1}{3\varepsilon^2}\left(1 + 3\varepsilon^2 + \sqrt{1 - 3\varepsilon^2}\right). \tag{4.2.45}$$

In this case we obtain a weak solution, namely the inverted coshoidal wave

$$z = z(\eta - 2j\eta_m), \quad (2j-1)\eta_m \leq \eta \leq (2j+1)\eta_m, \quad j = 0, \pm 1, \pm 2, \ldots, \tag{4.2.46}$$

where

$$\begin{aligned} z(\eta) &:= [z_3 - z_4 \tanh^2(\varepsilon\eta/2)] \cosh^2(\varepsilon\eta/2) \equiv \\ &\equiv \frac{1}{3\varepsilon^2}\left(1 + 3\varepsilon^2 - \sqrt{1 - 3\varepsilon^2}\cosh(\varepsilon\eta)\right) \end{aligned} \tag{4.2.47}$$

and

$$\eta_m = \frac{2}{\varepsilon}\tanh^{-1}\left(\sqrt{\frac{z_3}{z_4}}\right) \equiv \frac{1}{\varepsilon}\cosh^{-1}\left(\frac{3\varepsilon^2 + 1}{\sqrt{1 - 3\varepsilon^2}}\right); \tag{4.2.48}$$

see Fig. 4.7(a) for an example.

In the limit $\varepsilon^2 \to 0$, it is straightforward to show analytically that the inverted coshoidal-wave solution (4.2.46) reduces, as expected, to the inverted paraboidal-wave solution (3.2.18) of the VE for the case $v < 0$.

As ε^2 increases from 0 to $\frac{1}{3}$, the amplitude z_3 of the coshoidal wave increases from $\frac{3}{2}$ to 2, and its wavelength $\lambda := 2\eta_m$ increases from 6 to infinity.

4.2.2 f. $c = -1$, $\varepsilon^2 = 1/3$

In this case $0 < z_L < z_U$ with $f(0) = f(z_U)$. With $B_L < B < 0$ so that $0 < m < 1$, where $B_L = -1$, there are periodic well solutions to (4.2.25) given by (A.5) and (A.7), with wavelength given by (A.8); see Fig. 4.6(b) for an example. An alternative solution is given by (A.11) and (A.12); this is just the former solution phase-shifted by $\lambda/2$.

$B = 0$ corresponds to the limit $z_1 = z_2 = 0$ and $z_3 = z_4 = z_U = 2$. In this case neither (A.9) nor (A.14) is appropriate. Instead we consider (4.2.25) with $f(z) = z^2(2 - z)^2$ and note that the bound solution has $0 \leq z < 2$. On integrating (4.2.25) and setting $z = 0$ at $\eta = 0$ we obtain the weak solution

$$z = 2(1 - \exp[-|\eta|/\sqrt{3}\,]), \tag{4.2.49}$$

i.e. a single inverted peakon with amplitude 2; see Fig. 4.7(b).

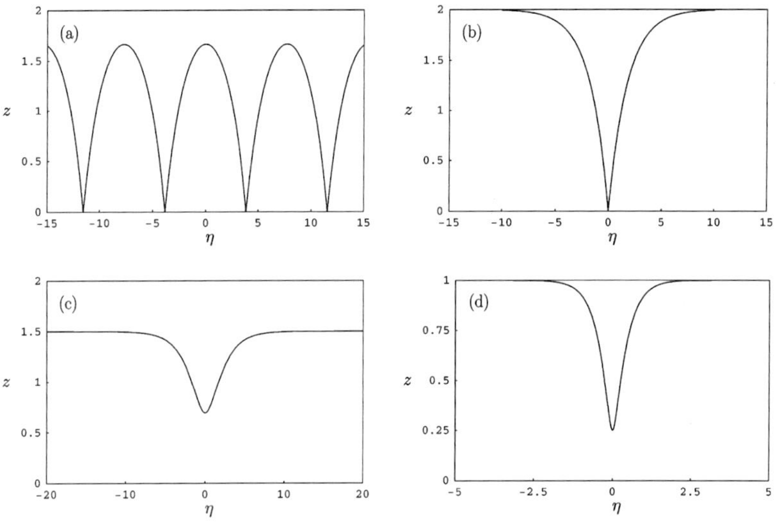

Figure 4.7: Solutions of the transformed DPE with $c = -1$ and $m = 1$: (a) $\varepsilon^2 = 1/4$, $B = 0$, $\lambda = 7.699$; (b) $\varepsilon^2 = 1/3$, $B = 0$; (c) $\varepsilon^2 = 1/2$, $B = B_U$; (d) $\varepsilon^2 = 8$, $B = B_L$.

4.2.2 g. $c = -1$, $1/3 < \varepsilon^2 < 1$

In this case $0 < z_L < z_U$ with $f(0) < f(z_U)$. For each value of ε^2 satisfying $1/3 < \varepsilon^2 < 1$ there are periodic well solutions to (4.2.25) given by (A.5) and (A.7) with $B_L < B < B_U$ so that $0 < m < 1$, and with wavelength given by (A.8); see Fig. 4.6(c) for an example.

$B = B_U$ corresponds to the limit $z_3 = z_4 = z_U$ so that $m = 1$, and then the solution is a well-like solitary wave given by (A.9) with $z_2 \leq z < z_U$ and

$$z_1 = \frac{1}{6\varepsilon^2}\left(-1 + 3\varepsilon^2 - \sqrt{2(9\varepsilon^4 - 1)}\right), \tag{4.2.50}$$

$$z_2 = \frac{1}{6\varepsilon^2}\left(-1 + 3\varepsilon^2 + \sqrt{2(9\varepsilon^4 - 1)}\right);$$

see Fig. 4.7(c) for an example. Note that $z_2 \to 1$ and $z_U \to 1$ as $\varepsilon^2 \to 1$, and that $z_2 \to 0$ and $z_U \to 2$ as $\varepsilon^2 \to 1/3$. As ε^2 decreases from 1 to 1/3, the amplitude $z_U - z_2$ of the solitary wave increases from 0 to 2.

4.2.2 h. $c = -1,\ \varepsilon^2 > 1$

In this case $0 < z_U < z_L$ with $f(0) < f(z_L)$. For each value of ε^2 satisfying $\varepsilon^2 > 1$ there are periodic well solutions to (4.2.25) given by (A.5) and (A.7) with $B_U < B < B_L$ so that $0 < m < 1$, and with wavelength given by (A.8); see Fig. 4.6(d) for an example.

$B = B_L$ corresponds to the limit $z_3 = z_4 = z_L = 1$ so that $m = 1$, and then the solution is a well-like solitary wave given by (A.9) with $z_2 \le z < z_L$ and

$$z_1 = \frac{1}{3\varepsilon^2}\left(1 - \sqrt{1 + 3\varepsilon^2}\right), \quad z_2 = \frac{1}{3\varepsilon^2}\left(1 + \sqrt{1 + 3\varepsilon^2}\right); \quad (4.2.51)$$

see Fig. 4.7(d) for an example. Note that $z_2 \to 1$ as $\varepsilon^2 \to 1$, and that $z_2 \to 0$ as $\varepsilon^2 \to \infty$. As ε^2 increases from 1 to infinity, the amplitude $1 - z_2$ of the solitary wave increases from 0 to 1.

4.2.3 Summary

We have found expressions for the travelling-wave solutions to the DPE that travel in the positive x-direction with speed v. These solutions depend, in effect, on two parameters A and m. In addition to the expected single peakon solution (with $A = 1$, $m = 1$) there are inverted loop-like ($A < 0$, $m = 1$) and hump-like ($1 < A < 9/8$, $m = 1$) solitary-wave solutions. For $0 < m < 1$ there are periodic inverted loop ($A < 0$) and periodic hump ($0 < A < 9/8$) solutions. For $m = 1$ and $0 < A < 1$ there are (periodic) coshoidal solutions. For each of the aforementioned solutions expressed with u as the dependent variable, there is a solution for u that is the mirror image in the x-axis and travels with the same speed but in the opposite direction.

We have also found expressions for the travelling-wave solutions to the transformed DPE. These solutions depend, in effect, on two parameters ε^2 and m, and also on the direction of propagation.

For propagation in the positive x-direction there are inverted loop-like ($1/3 < \varepsilon^2 < 1$, $m = 1$), kink-like ($\varepsilon^2 = 1/3$, $m = 1$) and loop-like ($0 < \varepsilon^2 < 1/3$, $m = 1$) solitary-wave solutions. For $0 < m < 1$ there are periodic hump ($\varepsilon^2 > 1$), periodic inverted-loop ($1/3 < \varepsilon^2 < 1$), periodic bell ($\varepsilon^2 = 1/3$) and periodic loop ($0 < \varepsilon^2 < 1/3$) solutions. For $m = 1$ and $\varepsilon^2 > 1$ there are (periodic) coshoidal solutions. In the limit $\varepsilon^2 \to 0$, the periodic loop solutions ($0 < m < 1$) and loop-like solitary-wave solutions ($m = 1$) to the VE are recovered.

For propagation in the negative x-direction there are inverted peakon ($\varepsilon^2 = 1/3$, $m = 1$) and well-like ($1/3 < \varepsilon^2 < 1$ and $\varepsilon^2 > 1$, $m = 1$) solitary-wave solutions. For $0 < m < 1$ there are periodic well ($0 < \varepsilon^2 < 1$ and $\varepsilon^2 > 1$) solutions. For $m = 1$ and $0 < \varepsilon^2 < 1/3$ there are (periodic) inverted coshoidal solutions. In the limit $\varepsilon^2 \to 0$, the periodic well solutions ($0 < m < 1$) and (periodic) inverted paraboidal solutions ($m = 1$) to the VE are recovered.

4.3 The Camassa–Holm equation

A classification o f t ravelling-wave s olutions o f t he C HE w as given in [66]. However, explicit solutions were given only for the solitary peakon and periodic peakon waves. Periodic smooth-hump waves and periodic cuspon waves were investigated numerically in [68].

Using a technique similar to the one we presented in [69] for the DPE, we obtain explicit travelling-wave solutions of the CHE (4.1.2)

$$u_t - u_{txx} + 3uu_x = 2u_x u_{xx} + uu_{xxx}$$

for both periodic and solitary smooth-hump, smooth-well, peakon, inverted-peakon, cuspon and inverted-cuspon waves [71].

4.3.1 Explicit travelling-wave solutions of the CHE

In terms of the new variables z and η for travelling-wave solutions as defined in Section 4.1.1, the CHE (4.1.2) has the form

$$zz_{\eta\eta\eta} + 2z_\eta z_{\eta\eta} - 3zz_\eta - 2cz_\eta = 0, \text{ where } c: = v/|v| = \pm 1. \quad (4.3.1)$$

After two integrations, (4.3.1) is reduced to

$$(zz_\eta)^2 = f(z), \quad (4.3.2)$$

where $f(z)$ is the polynomial given by (4.1.9) with $b = 2$. This polynomial can be written in terms of the roots of the equation $f(z) = 0$ as follows:

$$f(z): = z^4 + 2cz^3 + Az^2 + Bz \equiv (z - z_1)(z - z_2)(z_3 - z)(z_4 - z). \quad (4.3.3)$$

For the solutions of (4.3.2) that we are seeking, z_1, z_2, z_3 and z_4 are real constants with $z_1 \leq z_2 \leq z \leq z_3 \leq z_4$ and $z_1 + z_2 + z_3 + z_4 = -2c$.

From (4.3.3) it can be seen that one of z_1, z_2, z_3 and z_4 is always zero. We let the other three be q, r and s, where $s \leq r \leq q$ and $s = -q - r - 2c$. The types of solution to (4.3.2) may be categorized by an appropriate choice of the two parameters q and r. In [66, 68] the two parameters that were used, namely M and m in the notation of [66, 68], are equivalent to $q + c$ and $r + c$ respectively.

Equation (4.3.2) is of the same form as (A.1) in the Appendix to this Chapter (see Section 4.7) with $\varepsilon = 1$. Hence we can make use of the solutions given in the Appendix, but with $\varepsilon = 1$. Note that (4.3.2) is invariant under the transformation $z \to -z$, $c \to -c$; this corresponds to the transformation $u \to -u$, $v \to -v$ in (4.1.7). Here we will seek the family of solutions of (4.3.2) for which $v > 0$ in (4.1.7) and so, from here on in this Section, we will assume that $c = 1$.

4.3.1 a. $z_4 = 0$: Periodic smooth hump with $v > 0$

Suppose $z_4 = 0$ so that $z_1 = s$, $z_2 = r$ and $z_3 = q$. Consider the case $z_1 < z_2 < z_3 < 0$ so that

$$-q - r - 2c < r < q < 0. \tag{4.3.4}$$

(This is equivalent to the case considered numerically in Section 4.1 of [68].) The solution to (4.3.2) is a periodic hump given by (A.5) and (A.7), or (A.11) and (A.12), with $r \leq z \leq q$ and $0 < m < 1$; see Fig. 4.8 for an example given by (A.5) and (A.7).

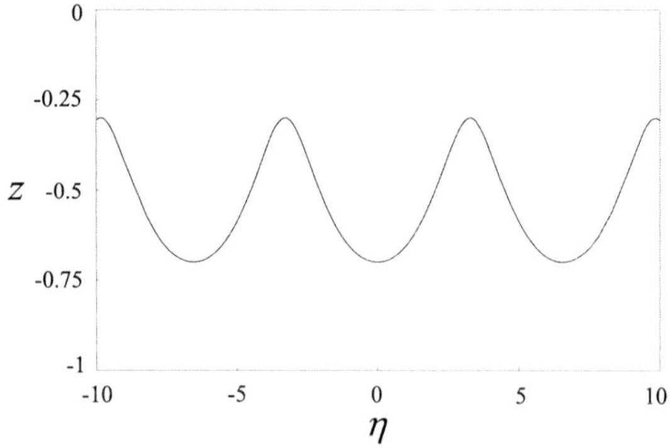

Figure 4.8: Periodic smooth hump solution of the CHE with $r = -0.7$, $q = -0.3$ and $v > 0$.

4.3.1 b. $z_4 = 0$: Solitary smooth hump with $v > 0$

In Section 4.3.1 a, consider the limit $z_1 = z_2$ so that we have $m = 1$ and $z_1 = z_2 < z_3 < 0$. In this case

$$-c < r < -\frac{2}{3}c, \quad q = -2(r + c). \tag{4.3.5}$$

The solution to (4.3.2) is a smooth-hump solitary wave given by (A.14) with $r < z \leq -2(r + c)$; see Fig. 4.9 for an example.

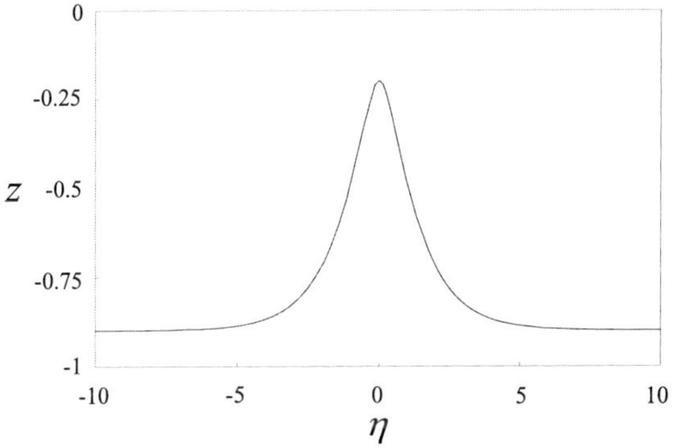

Figure 4.9: Solitary smooth hump solution of the CHE with $r = -0.9$, $q = -0.2$ and $v > 0$.

4.3.1 c. $z_4 = 0$: Periodic peakon with $v > 0$

In Section 4.3.1 a, consider the limit $z_3 = z_4$ so that we have $m = 1$ and $z_1 < z_2 < z_3 = 0$. In this case

$$-c < r < 0, \quad q = 0. \tag{4.3.6}$$

The solution to (4.3.2) is given by (A.9) and has $r \leq z \leq 0$. From this we can construct a weak solution, namely the periodic peakon wave given by

$$z = z(\eta - 2j\eta_m), \quad (2j-1)\eta_m \leq \eta \leq (2j+1)\eta_m, \quad j = 0, \pm 1, \pm 2, \ldots, \tag{4.3.7}$$

where

$$z(\eta) := [z_2 - z_1 \tanh^2(\eta/2)] \cosh^2(\eta/2) = -c + (r+c) \cosh \eta \quad (4.3.8)$$

and

$$\eta_m := 2 \tanh^{-1}\left(\sqrt{\frac{z_2}{z_1}}\right) = 2 \tanh^{-1}\left(\sqrt{\frac{-r}{r+2c}}\right); \quad (4.3.9)$$

see Fig. 4.10 for an example. The solution given by (4.3.7)–(4.3.9) is the spatially periodic solution of the CHE that has been dubbed a 'coshoidal wave' by Boyd [109].

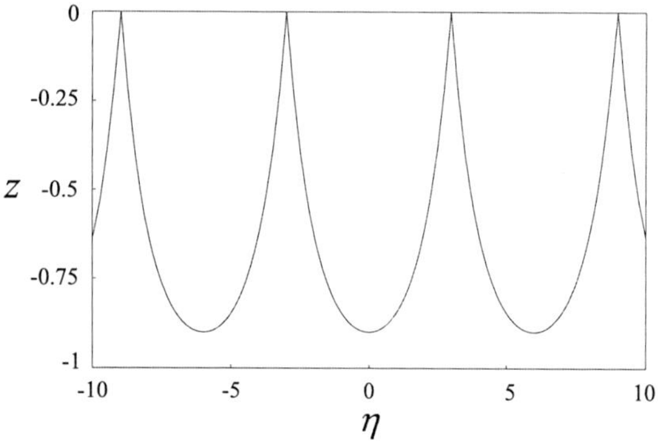

Figure 4.10: Periodic peakon solution of the CHE with $r = -0.9$, $q = 0$ and $v > 0$.

4.3.1 d. $z_4 = 0$: Solitary peakon with $v > 0$

In Section 4.3.1 a, consider the limit $z_1 = z_2$ and $z_3 = z_4$ so that we have $z_1 = z_2 < z_3 = 0$ and then

$$r = -c, \quad q = 0. \quad (4.3.10)$$

In this case neither (A.9) nor (A.14) is appropriate. Instead we consider (4.3.2) with $f(z) = z^2(z+c)^2$ and note that the bound solution has $-c < z \leq 0$. On integrating (4.3.2) and setting $z = 0$ at $\eta = 0$ we obtain the weak solution

$$z = c(e^{-|\eta|} - 1), \quad (4.3.11)$$

i.e. a solitary peakon with amplitude c; see Fig. 4.11.

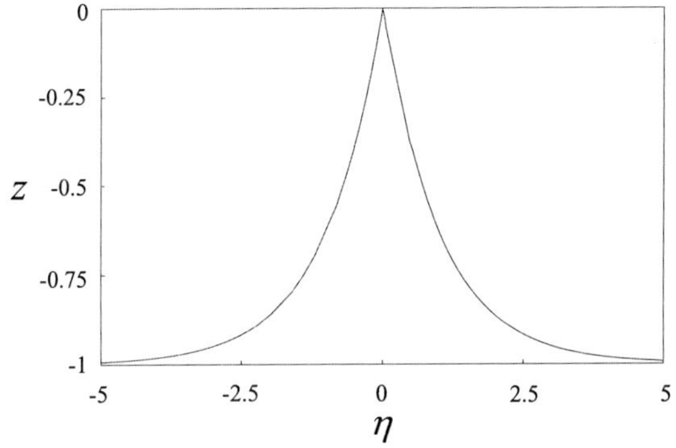

Figure 4.11: Solitary peakon solution of the CHE with $r = -1$, $q = 0$ and $v > 0$.

4.3.1 e. $z_3 = 0$: Periodic cuspon with $v > 0$

Suppose $z_3 = 0$ so that $z_1 = s$, $z_2 = r$ and $z_4 = q$. First let us consider the case $z_1 < z_2 < 0 < z_4$ so that

$$-q - r - 2c < r < 0 < q. \tag{4.3.12}$$

(This is equivalent to the case considered numerically in Section 4.2 of [68].) The solution to (4.3.2) is a periodic cuspon given by (A.5) and (A.7), or (A.11) and (A.12), with $r \leq z \leq 0$ and $0 < m < 1$; see Fig. 4.12 for an example given by (A.11) and (A.12).

4.3.1 f. $z_3 = 0$: Solitary cuspon with $v > 0$

In Section 4.3.1 e, consider the limit $z_1 = z_2$ so that we have $m = 1$ and $z_1 = z_2 < 0 < z_4$. In this case

$$r < -c, \quad q = -2(r + c). \tag{4.3.13}$$

The solution to (4.3.2) is a solitary cuspon given by (A.14) with $r < z \leq 0$; see Fig. 4.13 for an example.

4.3.1 g. $z_2 = 0$: Periodic inverted cuspon with $v > 0$

Suppose $z_2 = 0$ so that $z_1 = s$, $z_3 = r$ and $z_4 = q$. First let us consider the case $z_1 < 0 < z_3 < z_4$ so that

$$-q - r - 2c < 0 < r < q. \tag{4.3.14}$$

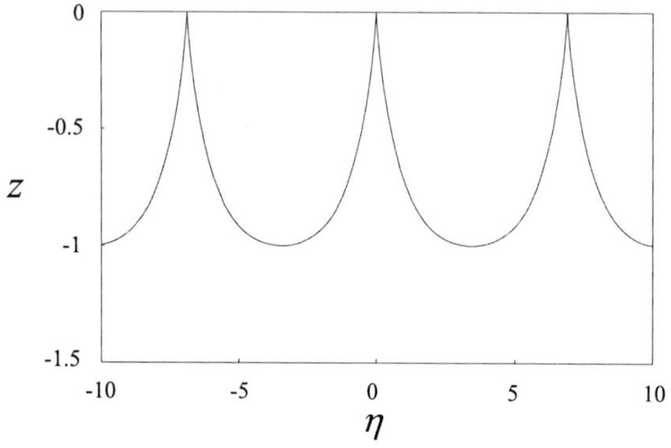

Figure 4.12: Periodic cuspon solution of the CHE with $r = -1$, $q = 0.1$ and $v > 0$.

The solution to (4.3.2) is a periodic inverted cuspon given by (A.5) and (A.7), or (A.11) and (A.12), with $0 \leq z \leq r$ and $0 < m < 1$; see Fig. 4.14 for an example given by (A.5) and (A.7).

4.3.1 h. $z_2 = 0$: Solitary inverted cuspon with $v > 0$

In Section 4.3.1 g, consider the limit $z_3 = z_4$ so that we have $m = 1$ and $z_1 < 0 < z_3 = z_4$. In this case

$$0 < r = q. \tag{4.3.15}$$

The solution to (4.3.2) is a solitary inverted cuspon given by (A.9) with $0 \leq z < r$; see Fig. 4.15 for an example.

4.3.1 i. $z_1 = 0$ and $v > 0$

In this case $z_2 + z_3 + z_4 > 0$ and so the condition $z_2 + z_3 + z_4 = -2c$ cannot be satisfied. Hence there are no solutions with $z_1 = 0$.

4.3.2 Further comments

In Sections 4.3.1 a – 4.3.1 h we have found explicit expressions for eight different travelling-wave solutions to the CHE that travel in the positive x-direction with speed v, i.e. with $v > 0$. These solutions depend on two parameters q and r. For each of the aforementioned solutions

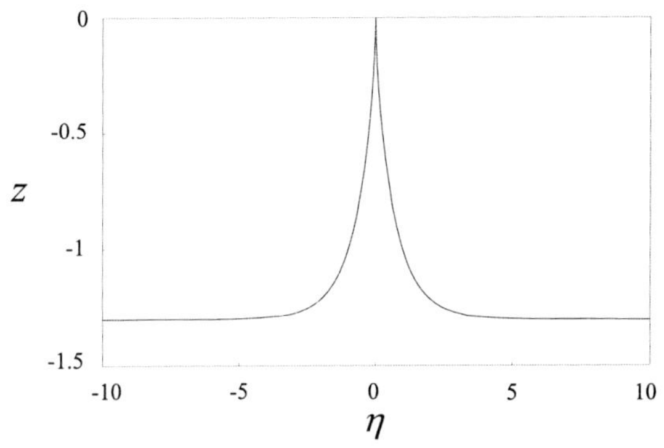

Figure 4.13: Solitary cuspon solution of the CHE with $r = -1.3$, $q = 0.6$ and $v > 0$.

expressed with u as the dependent variable, there is a solution for u that is the mirror image in the x-axis and travels with the same speed but in the opposite direction, i.e. with $v < 0$. For example, the mirror image of categories 4.3.1 b and 4.3.1 e respectively are solitary smooth wells with $v < 0$ and periodic inverted cuspons with $v < 0$.

In Theorem 1 in [66], Lenells categorized travelling-wave solutions to the CHE. His categories (a) – (d) correspond to our categories 4.3.1 a – 4.3.1 d. His category (e), i.e. periodic cuspons, correspond to our category 4.3.1 e, i.e. periodic cuspons with $v > 0$, together with the mirror image of our category 4.3.1 g, i.e. periodic cuspons with $v < 0$. His category (f), i.e. solitary cuspons, correspond to our category 4.3.1 f, i.e. solitary cuspons with $v > 0$, together with the mirror image of our category 4.3.1 h, i.e. solitary cuspons with $v < 0$. His categories (a′) – (f′) are the mirror images of his categories (a) – (f) respectively.

In Section 4.2 (see also [69]), we investigated the DPE. As is the case for the CHE, for $v > 0$ we found explicit expressions for smooth-hump and peakon solitary waves and their periodic equivalents. Unlike the CHE, for which we have found cuspon and inverted-cuspon solutions, we showed in Section 4.2 that the DPE has inverted loop-like solutions instead. However, it should be noted that it is possible to construct other explicit solutions for the DPE as composite waves by using the results in Section 4.2. Some examples are given in Appendix B in [71].

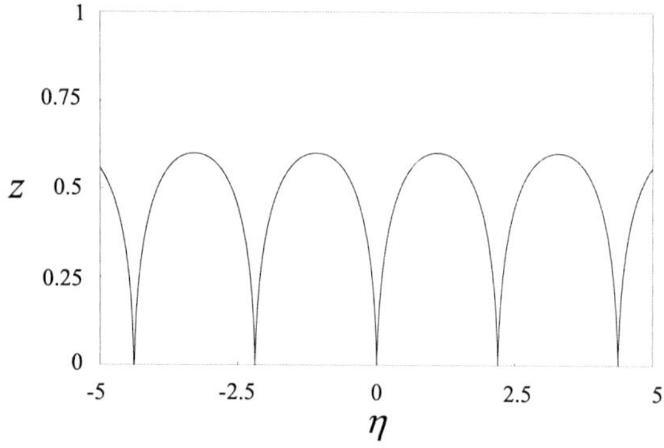

Figure 4.14: Periodic inverted cuspon solution of the CHE with $r = 0.6$, $q = 0.7$ and $v > 0$.

4.4 The generalized Degasperis–Procesi equation

4.4.1 The classification of travelling-wave solutions

Since the different polynomials in (4.2.3) and (4.3.3) can be written in the same form, as is shown by the right-hand sides in (4.2.3) and (4.3.3), we anticipate that there is a nonlinear equation which, for travelling-wave solutions, will reduce to

$$(zz_\eta)^2 = f(z), \tag{4.4.1}$$

with

$$f(z) = z^4 + 2cz^3 + Az^2 + Dz + B = (z - z_1)(z - z_2)(z - z_3)(z - z_4),$$

where A, B and D are real constants. It follows that this equation should be solvable in a way similar to that for the CHE and the DPE.

Let us consider the new nonlinear evolution equation [99]

$$(u_t+uu_x)^{b-1}(u_t-u_{txx}+(b+1)uu_x-bu_xu_{xx}-uu_{xxx})+\tfrac{1}{2}(2-b)D|v|^b u_x^b = 0. \tag{4.4.2}$$

Equation (4.4.2) generalises Eq. (4.1.1) due to the inclusion of an additional factor and an additional term. For travelling waves, Eq. (4.4.2)

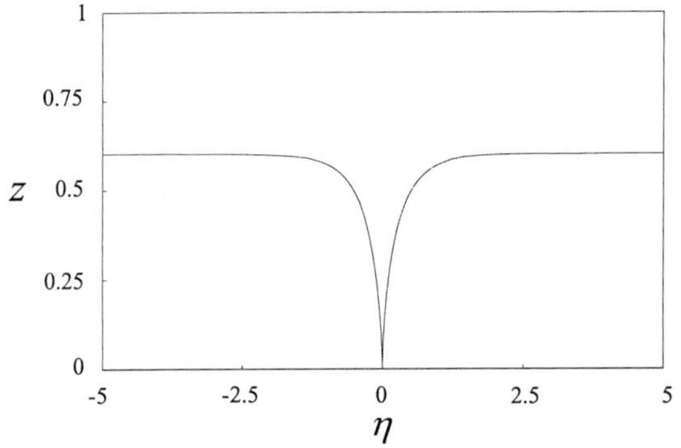

Figure 4.15: Solitary inverted cuspon solution of the CHE with $r = q = 0.6$ and $v > 0$.

in terms of the variables z and η as defined in Section 4.1.1 has the form

$$z^{b-1}(zz_{\eta\eta\eta}+bz_\eta z_{\eta\eta}-(b+1)zz_\eta-bcz_\eta)-\tfrac{1}{2}(2-b)Dz_\eta=0, \text{ with } c=\pm 1. \tag{4.4.3}$$

It can be seen that Eq. (4.4.3) generalises Eq. (4.1.8) due to the inclusion an additional factor z^{b-1} and an additional term $\frac{1}{2}(2-b)Dz_\eta$. After two integrations we get

$$(zz_\eta)^2 = f(z), \quad \text{with} \quad f(z) = z^4 + 2cz^3 + Az^2 + Dz^{4-b} + Bz^{3-b}. \tag{4.4.4}$$

With $b = 3$, Eq. (4.4.2) becomes

$$(u_t+uu_x)^2(u_t-u_{txx}+4uu_x-3u_xu_{xx}-uu_{xxx})-\tfrac{1}{2}D|v|^3u_x^3=0. \tag{4.4.5}$$

This will be referred to hereafter as the generalized Degasperis–Procesi equation (GDPE). Also, with $b = 3$, Eq. (4.4.3) becomes

$$z^2(zz_{\eta\eta\eta}+3z_\eta z_{\eta\eta}-4zz_\eta-3cz_\eta)+\tfrac{1}{2}Dz_\eta=0$$

and Eq. (4.4.4) becomes Eq. (4.4.1). Eq. (4.4.1) with $B = 0$ corresponds to the CHE for which $f(z)$ is given by (4.3.3); Eq. (4.4.1) with $D = 0$ corresponds to the DPE for which $f(z)$ is given by (4.2.3).

In principle, as the polynomial in (4.4.1) is a quartic, we can use the method of integration we applied to the CHE and the DPE to integrate the GDPE (4.4.5) and obtain travelling-wave solutions in the forms given by Eqs. (A.5) and (A.7), or (A.11) and (A.12), in the Appendix to this Chapter (see Section 4.7).

It is necessary to note that $f(z)$ in (4.4.1) involves three arbitrary constants A, B, D in contrast to $f(z)$ in (4.2.3) and (4.3.3) where there are only two constants. Hence, the GDPE should possess a wider variety of travelling-wave solutions than either the CHE or the DPE.

Since Eq. (4.4.1) is invariant under the transformation $z \to -z$, $c \to -c$, $D \to -D$, we need to consider only the case $c = 1$ (i.e. $v > 0$). Note that there is a restriction on the roots; they cannot be arbitrary because $z_1 + z_2 + z_3 + z_4 = -2c$ and they must be real.

In Table 4.1 we classify the different types of travelling-wave solutions of the GDPE (4.4.5) according to the disposition of the real roots of the polynomial $f(z)$. With distinct roots, the solutions are shown in the first column of Table 4.1 (Figs. 1.1–1.5). When $z_1 \neq z_2$ and $z_3 = z_4$, the solutions take the forms which are shown in the second column of Table 4.1 (Figs. 2.1–2.5). When $z_1 = z_2$ and $z_3 \neq z_4$ the solutions are shown in the third column of Table 4.1 (Figs. 3.1–3.3). Finally, in the fourth column of Table 4.1 (Figs. 4.1–4.3), there are the solutions with $z_1 = z_2$ and $z_3 = z_4$.

It should be noted that it is possible to construct other explicit solutions as composite waves [66] by using separate parts of the solutions from Table 4.1. Examples of this procedure have been given in Appendix B in [71]. In particular, the closed curve in Fig. 1.3b in Table 4.1 can be used to construct a periodic bell-like solution (see Fig. 4.5(c) in Section 4.2.2), while the two-valued solution in Fig. 4.3a and Fig. 4.3b in Table 4.1 can be used to construct a kink-like solution with infinite slope (see Fig. 4.6(c) in Section 4.2.2). Since these solutions are combined only from parts of the solutions we show in Table 4.1, such composite solutions are not presented in Table 4.1.

4.4.2 The graphical interpretation of the solutions

In this Section we suggest a graphical interpretation of the solutions from the Table 4.1. Let us consider a 3D spiral. It is shown in the colour black in the first column of Table 4.2. If we project the spiral perpendicularly to the spiral axis, then we will see the periodic hump given by the blue curve in Fig. 1.1 in Table 4.2. (All solutions are shown in the colour blue in Table 4.2.) At a specific projection angle to the spiral axis, the projection of the spiral will appear as a periodic cuspon (Fig. 1.2 in Table 4.2). Changing the angle between the direction of observation and the axis of the spiral, we can then see a periodic-loop solution (Fig. 1.3a in Table 4.2). In the exceptional case, when the observation takes place along the spiral axis, the spiral appears as a closed curve (Fig. 1.3b in Table 2). Thereafter the solutions are repeated in the reverse sequence: a periodic inverted loop solution (Fig. 1.3c in Table 4.2), a periodic inverted cuspon (Fig. 1.4 in Table 4.2), a periodic-hump solution (Fig. 1.5 in Table 4.2). Hence, we see all the solutions from the first column of Table 4.1.

To interpret the solutions from the second and third columns of Table 4.1, let us consider the black curves in the relevant columns of Table 4.2. These curves comprise one loop taken from a spiral. In the second column in Table 4.2, the upper part of the loop is extended, whereas in the third column the lower part is extended. At different projection angles for these curves on the plane, we observe a solitary smooth hump (Fig. 2.1 in Table 4.2), a hump-like solitary wave (Fig. 3.1 in Table 4.2), a periodic peakon (Fig. 2.2 in Table 4.2), a solitary cuspon (Fig. 3.2 in Table 2), a loop-like solitary wave (Figs. 2.3 and 3.3 in Table 4.2), a solitary inverted cuspon (Fig. 2.4 in Table 4.2), and an inverted hump (Fig. 2.5 in Table 4.2).

Finally, let us consider the 3D curve, which is shown in the fourth column of Table 4.2 in the colour black. It is none other than a half loop of a spiral with expanded upper and lower parts. This curve enables us to interpret the solutions (colour blue) from the fourth column. The projections give a kink-like solitary wave (Fig. 4.1 in Table 4.2), a single peakon solution (Fig. 4.2 in Table 4.2), and finally, a two-valued solution (Fig. 4.3 in Table 4.2).

Consequently, all the types of the solution from Table 4.1 are interpreted in Table 4.2.

4.4.3 Summary

We have suggested a new nonlinear evolution equation generalizing both the CHE and the DPE. This equation can be applied to describe shallow water waves, turbulent flows, and wave propagation in relaxing media. It can be integrated in a similar way to the CHE and the DPE in order to find travelling wave solutions. It turns out that the solutions of this new equation can be interpreted as the projection of a spiral on a plane at different projection angles to the axis of the spiral. The classification of the travelling wave solutions that we have presented in Section 4.4 may be of help in the understanding and description of the physical processes being investigated.

4.5 The Hirota–Satsuma-type 'shallow water wave' equation

We consider a Hirota–Satsuma-type 'shallow water wave' equation [97] of the form

$$U_{XXT}+pUU_T-qU_X\int_X^{\infty}U_T(X',T)dX'+\beta U_T+qU_X=0, \quad (4.5.1)$$

where $p\neq 0$, $q\neq 0$ and β are arbitrary constants.

Two special cases of (4.5.1) (with the rescaling $T\to -T/q$) have been studied in the literature. The case $p=2q$ and $\beta=-1$ was discussed by Ablowitz et al. [98] and was shown to be integrable by inverse scattering. This case, and the case $p=q$ with $\beta=-1$, were discussed by Hirota and Satsuma [100] and were shown to be integrable by using the Hirota bilinear technique.

By using the transformation (for details, see Section 5.1)

$$x=\theta(X,T)=T+\int_{-\infty}^{X}U(X',T)\,dX'+x_0,\;\; t=X,\;\; u(x,t)=U(X,T), \quad (4.5.2)$$

where x_0 is a constant, we obtain the following equation:

$$\frac{\partial}{\partial x}\left(\mathcal{D}^2u+\frac{1}{2}pu^2+\beta u\right)+q\mathcal{D}u=0, \quad (4.5.3)$$

where the operator $\mathcal{D}$ is defined by

$$\mathcal{D}:=\frac{\partial}{\partial t}+u\frac{\partial}{\partial x}. \quad (4.5.4)$$

We shall refer to the transformed form of equation (4.5.1), namely equation (4.5.3), as a transformed 'shallow water wave' equation.

As in [101], we refer to (4.5.3) with $p = q = 1$, and β an arbitrary non-zero constant, as a generalised VE (GVE), namely

$$\frac{\partial}{\partial x}\left(\mathcal{D}^2 u + \frac{1}{2}u^2 + \beta u\right) + \mathcal{D}u = 0 \tag{4.5.5}$$

or equivalently

$$\left(\frac{\partial u}{\partial x} + \mathcal{D}\right)\left(\frac{\partial}{\partial x}\mathcal{D}u + u + \beta\right) = 0. \tag{4.5.6}$$

As in [102], we refer to (4.5.3) with $p = 2q$, and $\beta \neq 0$, as a modified generalised Vakhnenko equation (mGVE), namely

$$\frac{\partial}{\partial x}\left(\mathcal{D}^2 u + qu^2 + \beta u\right) + q\mathcal{D}u = 0. \tag{4.5.7}$$

Note that if $\beta = 0$, Eq. (4.5.6) can be reduced to the VE given by (2.3.2) as expected, namely

$$\frac{\partial}{\partial x}\mathcal{D}u + u = 0.$$

4.5.1 An integrated form of the Eq. (4.5.3)

We find implicit periodic and solitary travelling-wave solutions of the transformed 'shallow water wave' equation (4.5.3), such that have the property that they reduce to the bounded solutions of the VE (2.3.2) for the appropriate choice of parameters, namely $p = q = 1$ and $\beta = 0$. This consideration leads us to seek solutions of the transformed 'shallow water wave' equation subject to the restrictions $p + q \neq 0$, $qv - \beta \neq 0$ [72].

In order to seek travelling-wave solutions of equation (4.5.3), it is convenient to introduce a new dependent variable z defined by

$$z = \frac{(p+q)u}{2|qv-\beta|} - c, \quad \text{where} \quad c{:}= \frac{qv-\beta}{|qv-\beta|} = \pm 1, \tag{4.5.8}$$

and to assume that z is an implicit or explicit function of η which is defined by

$$\eta = \frac{\chi}{|qv-\beta|^{1/2}}, \text{ where } \chi{:}= x - vt - x_0. \tag{4.5.9}$$

It is also convenient to introduce the variable ζ defined by the relation

$$\frac{d\eta}{d\zeta} = \frac{u - v}{|qv - \beta|} . \tag{4.5.10}$$

(Note that ζ is not a new spatial variable; it is the parameter in the parametric form of solution that we obtain eventually.) Then (4.5.3) becomes

$$z_{\zeta\zeta\zeta} + 2zz_\zeta + cz_\zeta = 0. \tag{4.5.11}$$

After one integration, (4.5.11) gives

$$z_{\zeta\zeta} + z^2 + cz = B, \tag{4.5.12}$$

where B is a constant of integration.

We impose the requirement that, for $p = q = 1$ and $\beta = 0$, the solutions that we seek reduce to the corresponding solutions of the VE. It turns out that Eq. (4.5.12) with $B = 0$ reduces to the corresponding relation for the VE. Accordingly we set $B = 0$ from here on.

With $B = 0$, equation (4.5.12) can be integrated once more to give

$$z_\zeta^2 = f(z) := -\tfrac{2}{3}z^3 - cz^2 + \tfrac{1}{3}c^3 A, \tag{4.5.13}$$

where A is a real constant. Equation (4.5.13) is equivalent to (3.2.4) which arises as one of the differential equations in solving the VE.

4.5.2 Travelling-wave solutions of the Eq. (4.5.3)

The bounded solutions of equation (4.5.13) that we seek are such that $z_1 \le z_2 \le z \le z_3$, where z_1, z_2 and z_3 are the three real roots of $f(z) = 0$. In equations (A.2)–(A.4) and (A.6) in the Appendix in Chapter 3 (Section 3.4), we gave expressions for these roots and $m := (z_3 - z_2)/(z_3 - z_1)$ in terms of an angle θ. By eliminating θ, we obtain z_1, z_2 and z_3 in terms of m, namely

$$z_1 = -\frac{c}{2} + \frac{m - 2}{2\sqrt{m^2 - m + 1}}, \tag{4.5.14}$$

$$z_2 = -\frac{c}{2} + \frac{1 - 2m}{2\sqrt{m^2 - m + 1}}, \tag{4.5.15}$$

$$z_3 = -\frac{c}{2} + \frac{1 + m}{2\sqrt{m^2 - m + 1}}, \tag{4.5.16}$$

where $0 \leq m \leq 1$. As in obtaining (3.2.10), we may integrate equation (4.5.13) by using result 236.00 in [44] to obtain

$$z = z_3 - (z_3 - z_2)\,\mathrm{sn}^2(w|m), \quad \text{where} \quad w = \sqrt{\frac{z_3 - z_1}{6}}\zeta. \tag{4.5.17}$$

Result 310.02 in [44] leads to

$$\int z dw = z_1 w + (z_3 - z_1)E(w|m) + \text{const}\,. \tag{4.5.18}$$

As in (3.2.10), $\mathrm{sn}(w|m)$ in (4.5.17) is a Jacobian elliptic function; as in (3.2.11), $E(w|m)$ in (4.5.18) is the incomplete elliptic integral of the second kind.

In view of the definition of c in (4.5.8), it is convenient to let

$$qv - \beta = 4c\kappa^2, \tag{4.5.19}$$

where κ is a positive constant. It is also convenient to define the positive constant κ by

$$\kappa^2 = k^2\sqrt{m^2 - m + 1}\,. \tag{4.5.20}$$

By using (4.5.14)–(4.5.20) in (4.5.8)–(4.5.10), we obtain

$$u = \frac{4k^2}{p+q}\left[m + 1 + c\sqrt{m^2 - m + 1} - 3m\,\mathrm{sn}^2(w|m)\right]. \tag{4.5.21}$$

$$\chi = \frac{4k^2}{p+q}\left[(m - 2 + c\sqrt{m^2 - m + 1})w + 3E(w|m)\right] - \frac{(\beta + 4c\kappa^2)w}{kq}. \tag{4.5.22}$$

The travelling-wave solution to the transformed 'shallow water wave' equation (4.5.3) is given in parametric form by (4.5.21) and (4.5.22) with w as the parameter, so that u is an implicit function of χ. With respect to w, u in (4.5.21) is periodic with period $2K(m)$, where $K(m)$ is the complete elliptic integral of the first kind. It follows from (4.5.22) that the wavelength λ of u regarded as an implicit function of χ is

$$\begin{aligned}\lambda = \Big|&\frac{8k}{p+q}[(m - 2 + c\sqrt{m^2 - m + 1})K(m) + 3E(m)] - \\ &- \frac{2(\beta + 4c\kappa^2)K(m)}{kq}\Big|,\end{aligned} \tag{4.5.23}$$

where $E(m)$ is the complete elliptic integral of the second kind.

When $m = 1$, the solution given by (4.5.21) and (4.5.22) becomes

$$u = \frac{4k^2}{p+q}\left[2 + c - 3\tanh^2 w\right], \tag{4.5.24}$$

$$\chi = \frac{4k}{p+q}\left[(c-1)w + 3\tanh w\right] - \frac{(\beta + 4ck^2)w}{kq}. \tag{4.5.25}$$

4.5.3 Examples

We illustrate the results in Section 4.5.2 by considering two examples.

Example 1

We consider the simplest case, namely the VE (for which $p = q = 1$ and $\beta = 0$). In this case, with $c = 1$, we have $v > 0$. Analysis shows that the solution comprises periodic upright loops for $0 < m < 1$ and a solitary upright loop for $m = 1$. On the other hand, for $c = -1$, we have $v < 0$. As result we obtain that the solution comprises periodic smooth humps for $0 < m < 1$ and a periodic corner-wave for $m = 1$. These are the results first given in [41, 42] (see also Section 3.2).

Example 2

We consider the GVE (4.5.5) (for which $p = q = 1$) with $c = 1$ and arbitrary β. (Note that the particular case for which $\beta = 0$ is just the VE with $c = 1$ as discussed in Example 1.) To illustrate the results in Section 4.5.2, we let $k = 1$ and consider the cases $m = 0.5$ and $m = 1$ separately.

Figs. 4.16–4.22 correspond to seven choices of β with $k = 1$ and $m = 0.5$. In each figure, χ is plotted as an explicit function of w, and u is plotted as an implicit function of χ. In Figs. 4.16–4.21, the wave profile is periodic. In Fig. 4.22, the solution is a closed curve in the (χ, u) plane. Fig. 4.23 illustrates the corresponding composite solution.

Figs. 4.24–4.28 correspond to five choices of β with $k = 1$ and $m = 1$. In Figs. 4.24–4.27, the wave profile is a solitary wave. In Fig. 4.28, the solution is a solitary wave with compact support. Fig. 4.29 illustrates the corresponding composite solution comprising periodic corner waves. The solitary waves for $\beta/k^2 \neq -4$ with $c = 1$, i.e. $v \neq 0$,

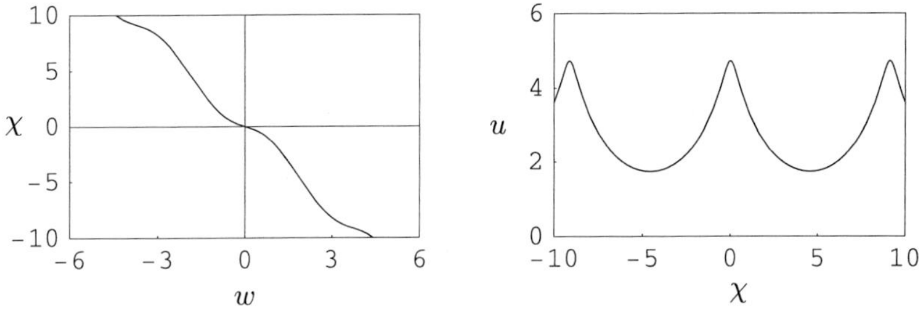

Figure 4.16: GVE with $\beta = 2.1$: The wave profile comprises periodic smooth humps with $v > 0$.

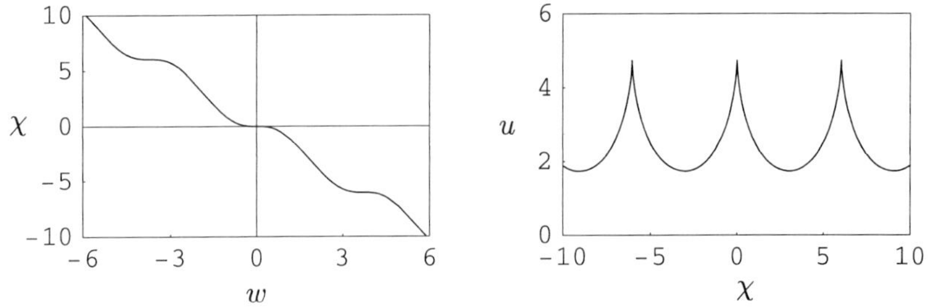

Figure 4.17: GVE with $\beta \simeq 1.27$: The wave profile comprises periodic cuspons with $v > 0$.

coincide with the single soliton solutions given in [101] for the GVE as derived via Hirota's method and illustrated in Fig. 4 in [101]. In [101] it was shown that Hirota's method fails when $v = 0$. Now we know that, in this case, the solution is a solitary wave with compact support or a periodic corner wave.

The multi-valued solution illustrated in Fig. 4.26 may be regarded as a composite solution of three single-valued solutions, but the three single-valued solutions may also be combined in different ways so as to give a variety of composite single-valued solutions. One of these possibilities is illustrated in Fig. 4.30. Note that, in this solution, η is a monotonic increasing function of the parameter w. The corresponding periodic single-valued composite solution may be constructed from the single-valued solutions making up the multi-valued waves in Figs. 4.18 and 4.19.

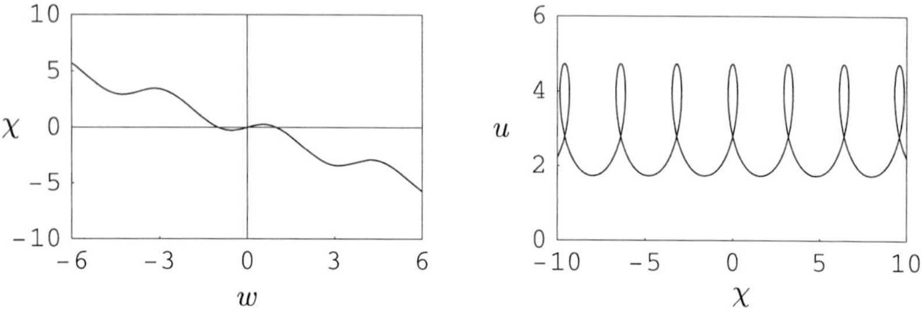

Figure 4.18: GVE with $\beta = 0.5$: The wave profile comprises periodic loops with $v > 0$.

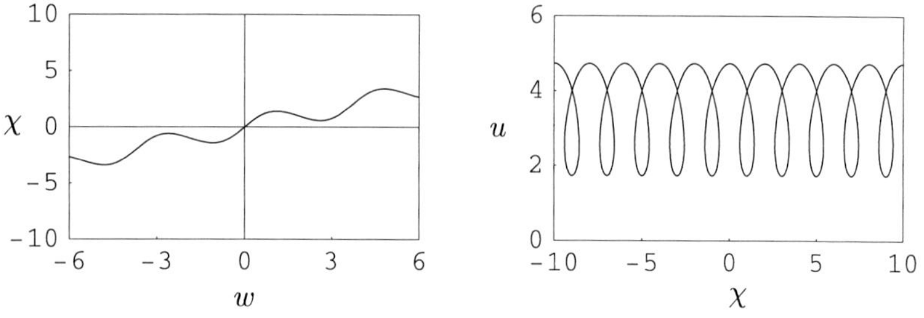

Figure 4.19: GVE with $\beta = -0.9$: The wave profile comprises periodic inverted loops with $v > 0$.

4.6 The short-pulse equation

The short-pulse equation (SPE), namely

$$u_{xt} = u + \frac{1}{6}(u^3)_{xx} \tag{4.6.1}$$

models the propagation of ultra-short light pulses in silica optical fibres [104].

In [105] it was shown that the SPE has a Lax pair that is of the Wadati–Konno–Ichikawa-type (see [106], for example). Because of this result, it is not surprising that the SPE has a loop-soliton solution. This solution was found in [107] together with several other forms of solution.

In passing we note that there several other equations that have

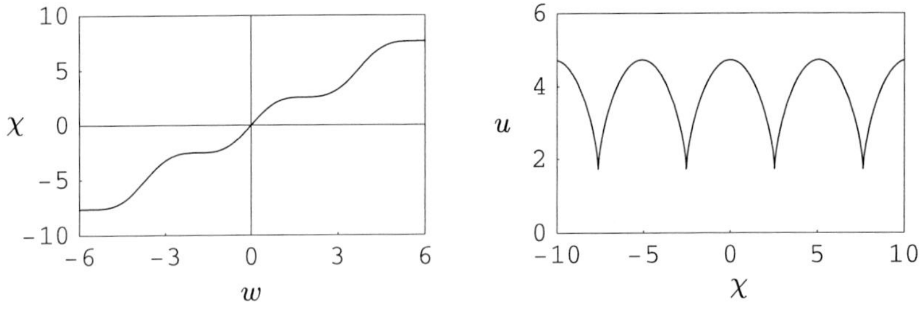

Figure 4.20: GVE with $\beta \simeq -1.73$: The wave profile comprises periodic inverted cuspons with $v > 0$.

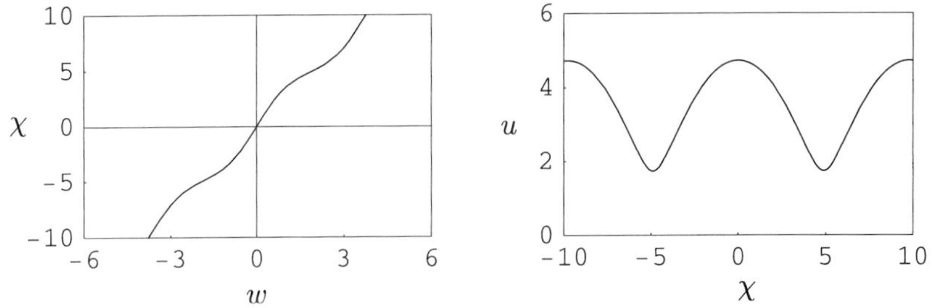

Figure 4.21: GVE with $\beta = -3$: The wave profile comprises periodic smooth humps with $v > 0$.

loop-soliton solutions; in [90] we gave a list of references in which some of these equations are discussed. We have presented two more such equations, namely the generalized Vakhnenko equation [101] and the modified generalized Vakhnenko equation [102] (see also Section 4.5).

In [105] it was shown that the SPE and the sine-Gordon equation (SGE) are equivalent to one another through a chain of transformations. In [107] various known solutions of the SGE were used to generate solutions to the SPE. The kink solution to the SGE leads to a travelling-wave solution of the SPE in the form of a loop soliton. The two-kink and kink-antikink solutions of the SGE also lead to multi-valued solutions of the SPE but they are not travelling waves. The breather solution to the SGE leads to a wave packet solution of the SPE. In the context of light pulses, the latter is the physically relevant solution.

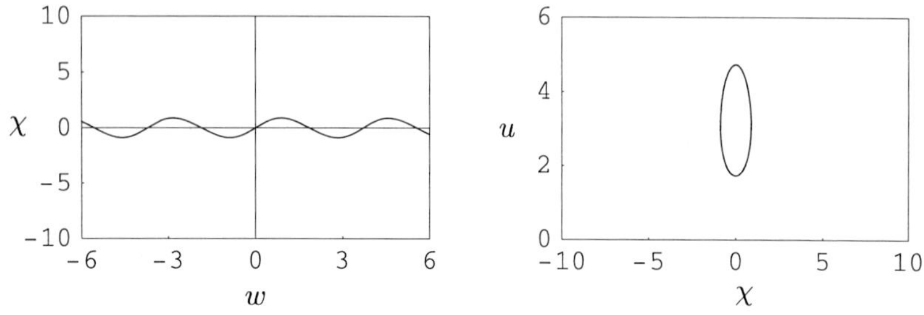

Figure 4.22: GVE with $\beta = -0.36$: The solution for u is a closed curve.

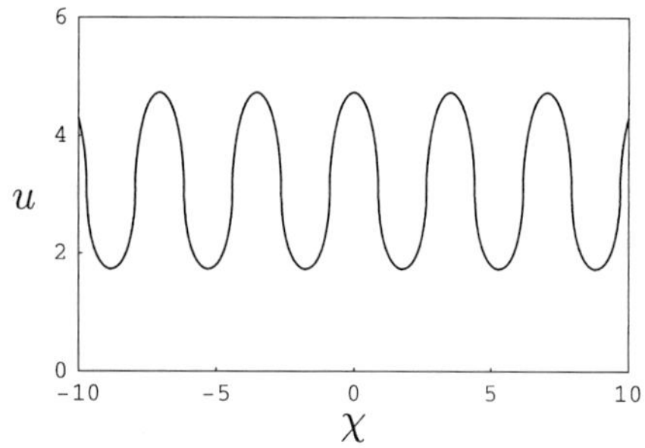

Figure 4.23: A composite solution for the GVE corresponding to Fig. 4.22. The wave profile comprises periodic bells.

We complement the work in [107] by presenting other travelling-wave solutions to the SPE [103]. The solution method is similar to the one that we have used previously to find periodic and solitary travelling-wave solutions to the Degasperis–Procesi equation [69] (see also Section 4.2), the Camassa–Holm equation [71] (see also Section 4.3) and the Vakhnenko equation [41, 42] (see also Section 3.2).

4.6.1 The direct method of integration

In order to seek travelling-wave solutions to the SPE it is convenient to introduce new variables z and η defined by

$$z = u/|v|^{1/2}, \quad \eta = (x - vt - x_0)/|v|^{1/2}, \tag{4.6.2}$$

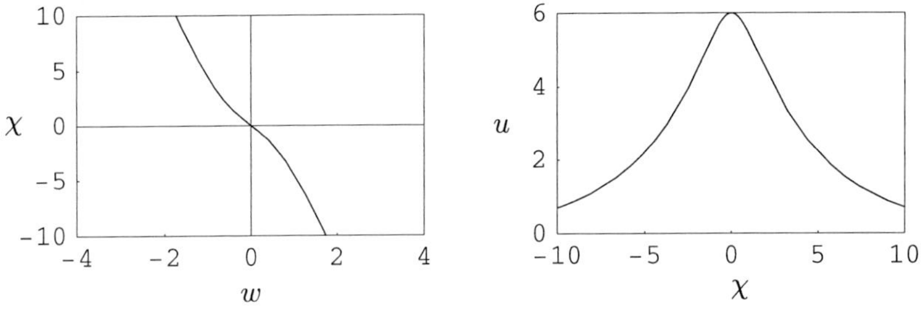

Figure 4.24: GVE with $\beta = 5$: The wave profile is a solitary smooth hump with $v > 0$.

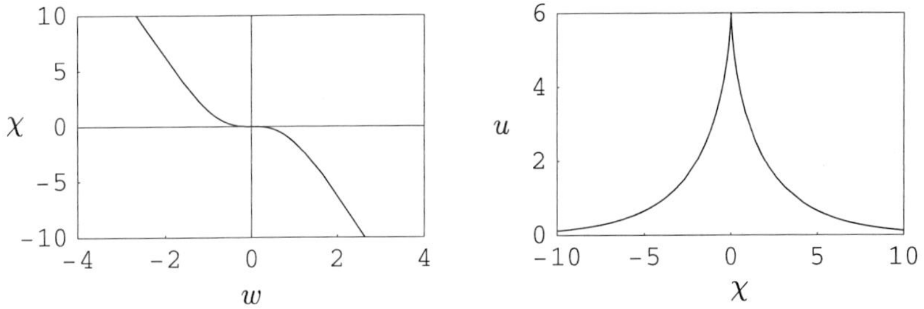

Figure 4.25: GVE with $\beta = 2$: The wave profile is a solitary cuspon with $v > 0$.

where z is an implicit or explicit function of η, and $v \neq 0$. In this case (4.6.1) becomes

$$(z^2 + 2c)z_{\eta\eta} + 2z(1 + z_\eta^2) = 0, \text{ where, } c: = v/|v| = \pm 1. \quad (4.6.3)$$

After one integration (4.6.3) gives

$$z_\zeta^2 = f(z), \quad (4.6.4)$$

where

$$f(z): = B^2 - (z^2 + 2c)^2, \quad (4.6.5)$$

B is a real positive constant and ζ is defined by

$$\frac{\mathrm{d}\eta}{\mathrm{d}\zeta} = z^2 + 2c. \quad (4.6.6)$$

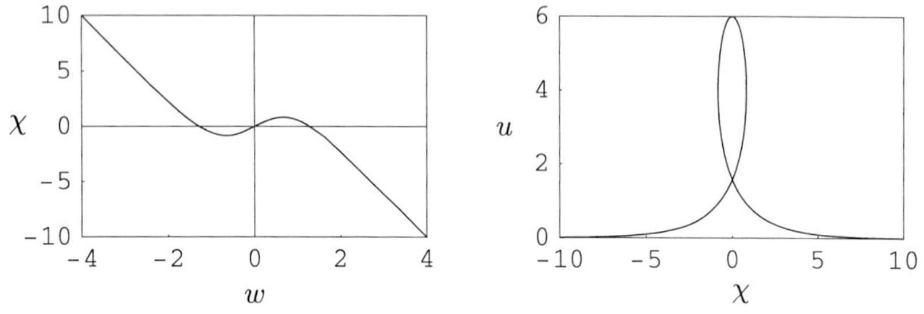

Figure 4.26: GVE with $\beta = 0$: The wave profile is a solitary loop with $v > 0$.

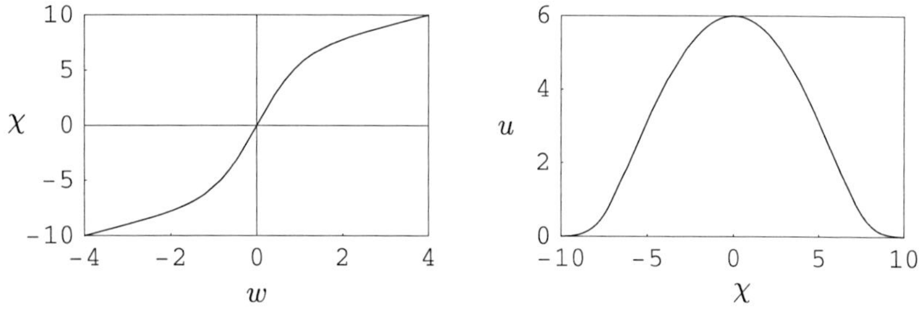

Figure 4.27: GVE with $\beta = -5$: The wave profile is a solitary smooth hump with $v < 0$.

We note that (4.6.3) is invariant under the transformation $z \to -z$. Also, when $c = -1$, (4.6.6) indicates that the solutions will have infinite slope when $z^2 = 2$.

4.6.2 Exact travelling-wave solutions of the SPE

For each choice of c, the possible types of travelling-wave solution of the SPE depend on the value, or range of values, of B. The types may be classified as described in Sections 4.6.2 a–d as follows:

4.6.2 a. $c = 1, \ B > 2$

In this case (4.6.5) may be written

$$f(z) := (a^2 + z^2)(b^2 - z^2), \text{ where } a^2 = B + 2, b^2 = B - 2. \quad (4.6.7)$$

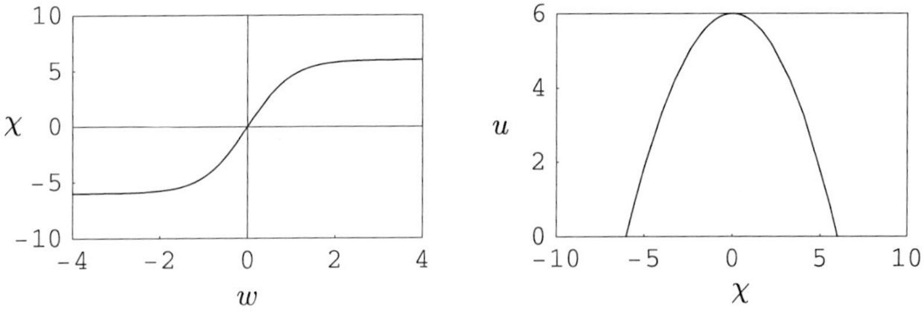

Figure 4.28: GVE with $\beta = -4$: The solution for u is a solitary wave with compact support and $v = 0$.

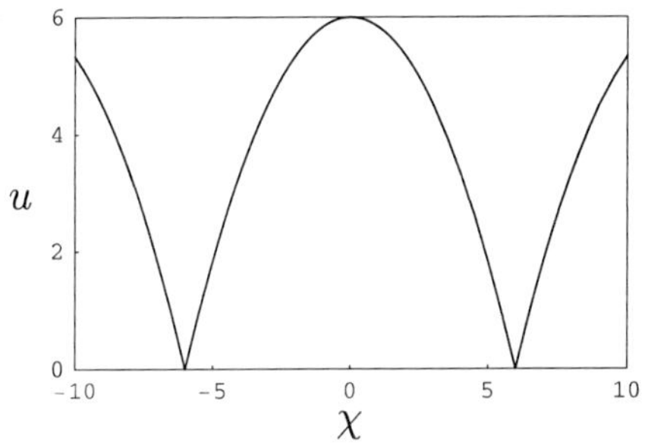

Figure 4.29: A composite solution of the GVE corresponding to Fig. 4.28. The wave profile comprises periodic corner waves.

The bounded solutions to (4.6.3) are such that $-b \leq z \leq b$. By using results 213.00 and 310.02 from [44] to integrate (4.6.4) and (4.6.6), we find that these solutions are given in parametric form by

$$z = \pm\sqrt{\frac{4m}{1-2m}}\,\mathrm{cn}(w|m), \quad \eta = \frac{[-w + 2E(w|m)]}{\sqrt{1-2m}}, \tag{4.6.8}$$

where $w(= 2\zeta/\sqrt{1-2m})$ is the parameter and

$$m = \frac{B-2}{2B} \quad \text{so that} \quad 0 < m < \frac{1}{2}.$$

In (4.6.8), $\mathrm{cn}(w|m)$ is a Jacobian elliptic function and $E(w|m)$ is the incomplete elliptic integral of the second kind.

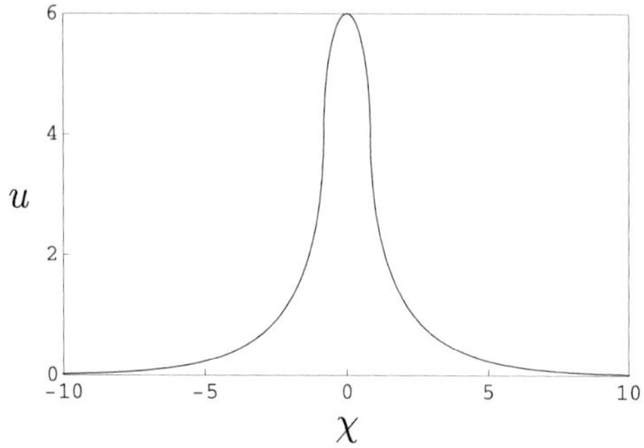

Figure 4.30: A single-valued composite solution of the GVE corresponding to Fig. 4.26.

The solution (4.6.8) is a periodic hump or a periodic well corresponding respectively to the upper or lower choice of sign in (4.6.8). The wavelength of these solutions is

$$\lambda = \frac{4|-K(m)+2E(m)|}{\sqrt{1-2m}}, \tag{4.6.9}$$

where $K(m)$ and $E(m)$ are the complete elliptic integrals of the first and second kind, respectively. Note that λ increases from 2π to infinity as m increases from 0 to 0.5. An example of the periodic hump solution is illustrated in Fig. 4.31.

4.6.2 b. $c = -1$, $B = 2$

In this case (4.6.5) may be written

$$f(z) := z^2(4 - z^2). \tag{4.6.10}$$

The bounded solutions to (4.6.3) are such that $0 \leq z^2 \leq 4$. Straightforward integration of (4.6.4) and (4.6.6) gives

$$z = \pm 2\,\mathrm{sech}\, w, \quad \eta = -w + 2\tanh w, \tag{4.6.11}$$

where $w(= 2\zeta)$ is the parameter. The solution (4.6.11) is either an upright or an inverted solitary loop corresponding respectively to the upper or lower choice of sign in (4.6.11). The upright solitary loop solution is illustrated in Fig. 4.32. With the upper choice of sign, (4.6.11) is equivalent to Eq. (13) in [107]. The latter solution was derived from a transformation of the kink solution of the SGE.

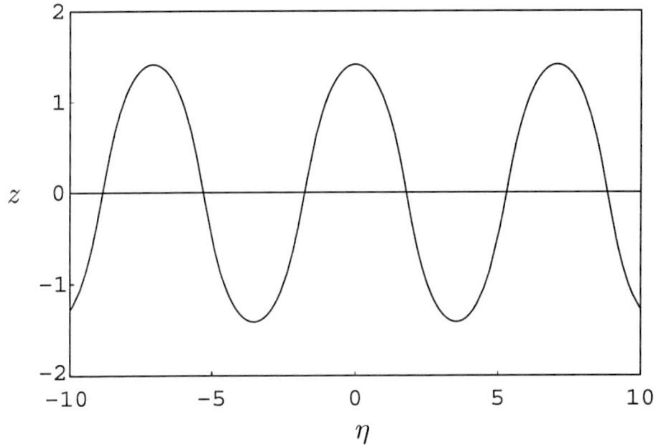

Figure 4.31: A typical periodic-hump solution of the SPE with $c = 1$ for which $B = 4$ so that $m = 0.25$ and $\lambda = 7.0664$.

4.6.2 c. $c = -1,\ 0 < B < 2$

In this case (4.6.5) may be written

$$f(z) := (a^2 - z^2)(z^2 - b^2), \text{ where } a^2 = 2+B,\ b^2 = 2-B. \quad (4.6.12)$$

The bounded solutions to (4.6.3) are such that $b^2 \leq z^2 \leq a^2$. By using results 218.00 and 310.02 from [44] to integrate (4.6.4) and (4.6.6), we find that these solutions are given in parametric form by

$$z = \pm \frac{2}{\sqrt{2-m}}\,\mathrm{dn}(w|m), \quad \eta = \frac{[-(2-m)w + 2E(w|m)]}{\sqrt{2-m}}, \quad (4.6.13)$$

where $w(= 2\zeta/\sqrt{2-m})$ is the parameter and

$$m = \frac{2B}{B+2} \quad \text{so that} \quad 0 < m < 1.$$

The solution (4.6.13) is either periodic upright loops or periodic inverted loops corresponding to the upper or lower choice of sign in (4.6.13), respectively. The wavelength of these solutions is

$$\lambda = \frac{2|-(2-m)K(m) + 2E(m)|}{\sqrt{2-m}}. \quad (4.6.14)$$

Note that λ increases from zero to infinity as m increases from 0 to 1. An example of the periodic upright loop solution is illustrated

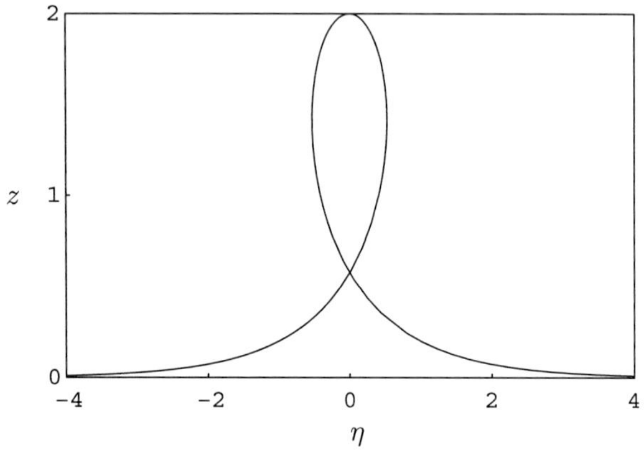

Figure 4.32: The solitary upright-loop solution of the SPE with $c = -1$ for which $B = 2$ so that $m = 1$.

in Fig. 4.33. In the limit $B \to 2$ (so that $m \to 1$), (4.6.13) becomes (4.6.11), i.e. the periodic loops degenerate to a solitary loop.

4.6.2 d. $c = -1$, $B > 2$

In this case (4.6.5) may be written

$$f(z) := (a^2+z^2)(b^2-z^2), \text{ where } a^2 = B-2,\ b^2 = B+2. \quad (4.6.15)$$

The bounded solutions to (4.6.3) are such that $-b \leq z \leq b$. By using results 213.00 and 310.02 from [44] to integrate (4.6.4) and (4.6.6), we find that these solutions are given in parametric form by

$$z = \pm\sqrt{\frac{4m}{2m-1}}\,\mathrm{cn}(w|m), \quad \eta = \frac{[-w + 2E(w|m)]}{\sqrt{2m-1}}, \quad (4.6.16)$$

where $w(= 2\zeta/\sqrt{2m-1})$ is the parameter and

$$m = \frac{B+2}{2B} \quad \text{so that} \quad \frac{1}{2} < m < 1.$$

The solution (4.6.16) is periodic with wavelength

$$\lambda = \frac{4|-K(m) + 2E(m)|}{\sqrt{2m-1}}. \quad (4.6.17)$$

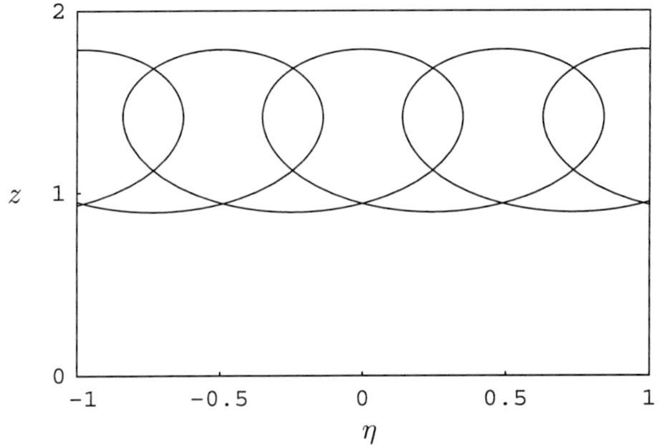

Figure 4.33: A typical periodic upright-loop solution of the SPE with $c = -1$ for which $B = 1.2$ so that $m = 0.75$ and $\lambda = 0.48931$.

Note that $\lambda = 0$ at $m = m_c = 0.826115$ and that λ increases from zero to infinity as m decreases from m_c to 0.5, and as m increases from m_c to 1. In the limit $B \to 2$ (so that $m \to 1$), (4.6.16) becomes (4.6.11).

When $m = m_c$ the solution (4.6.16) is a 'figure-of-eight' centered on the origin in the (η, z) plane as shown in Fig. 4.34. For $0.5 < m < m_c$ the solution comprises alternating upright and inverted 'bells', whereas for $m_c < m < 1$ the solution comprises alternating upright and inverted loops. Examples corresponding to the upper choice of sign in (4.6.16) are illustrated in Figs. 4.35 and 4.36, respectively.

For $m = m_c$, periodic composite solutions of (4.6.3) may be constructed. (The notion of composite waves is discussed in [66], for example.) Fig. 4.37 shows the periodic composite solution given in parametric form by

$$z = z(w), \tag{4.6.18}$$

$$\eta = \begin{cases} \eta(w) + 8j\eta(w_1), & -w_1 + 4jK \le w < w_1 + 4jK, \\ \eta(w) + (6+8j)\eta(w_1), & w_1 + 4jK \le w < -w_1 + (2+4j)K, \\ \eta(w) + (4+8j)\eta(w_1), & -w_1 + (2+4j)K \le w < w_1 + (2+4j)K, \\ \eta(w) + (2+8j)\eta(w_1), & w_1 + (2+4j)K \le w < -w_1 + (4+4j)K, \end{cases}$$

where $K := K(m_c)$, $z(w)$ and $\eta(w)$ are given by (4.6.16) with the upper choice of sign and $m = m_c$, w_1 is the root of $z(w) = \sqrt{2}$ such

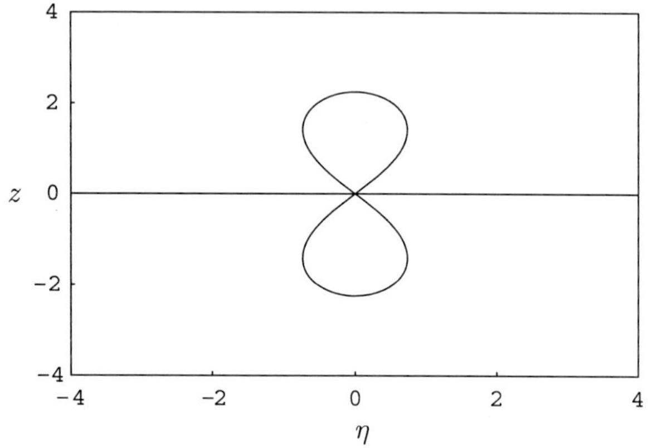

Figure 4.34: The 'figure-of-eight' solution of the SPE with $c = -1$ and $m = m_c = 0.826115$ so that $B = 3.06641$ and $\lambda = 0$.

that $0 < w_1 < K$, and $j = 0, \pm 1, \pm 2, \ldots,$. The wavelength of this solution is $\lambda = 8\eta(w_1)$. The construction of the solution (4.6.18) is similar to the example in Section 4.2.2 c and illustrated in Fig. 4.4(c).

4.6.3 Concluding remarks

We have found exact expressions for travelling-wave solutions to the SPE. For each of the two choices of c these solutions depend on one parameter B. There is one type of periodic-wave solution that propagates in the positive x–direction. There are several types of periodic-wave solution and one type of solitary-wave solution that propagate in the negative x–direction; there are places on the wave profile of these solutions where the slope goes infinite. Each solution has a corresponding solution that propagates in the same direction and is a mirror image in the x–axis.

4.7 Appendix

Here we consider solutions to

$$(zz_\eta)^2 = \varepsilon^2 f(z), \tag{A.1}$$

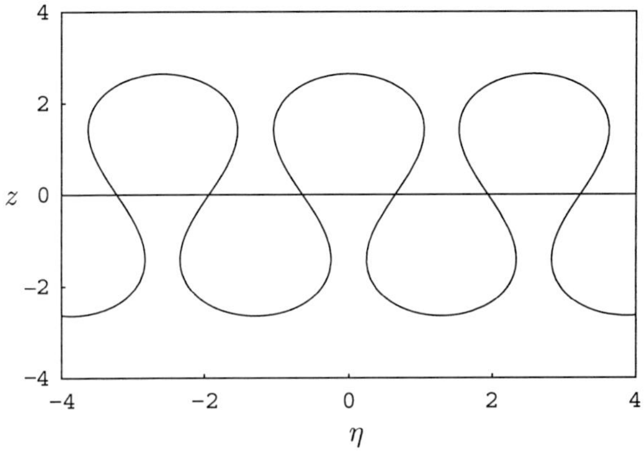

Figure 4.35: A typical periodic solution of the SPE with $c = -1$ comprising alternating upright bells and inverted bells for which $B = 5$ so that $m = 0.7$ and $\lambda = 2.58028$.

where

$$f(z) := (z - z_1)(z - z_2)(z_3 - z)(z_4 - z); \tag{A.2}$$

for the solutions that we are seeking, z_1, z_2, z_3 and z_4 are real constants with $z_1 \leq z_2 \leq z \leq z_3 \leq z_4$.

Following [42] we introduce ζ defined by

$$\frac{d\eta}{d\zeta} = \frac{z}{\varepsilon} \tag{A.3}$$

so that (A.1) becomes

$$z_\zeta^2 = f(z). \tag{A.4}$$

The solutions to (A.4) are found by integration, where the interval of integration is between the roots z_2 and z_3. (A.4) has two possible forms of solution.

The first form of solution of (A.4) is found using result 254.00 in [44]. It is

$$z = \frac{z_2 - z_1 n \,\mathrm{sn}^2(w|m)}{1 - n \,\mathrm{sn}^2(w|m)} \quad \text{with} \quad n = \frac{z_3 - z_2}{z_3 - z_1}, \tag{A.5}$$

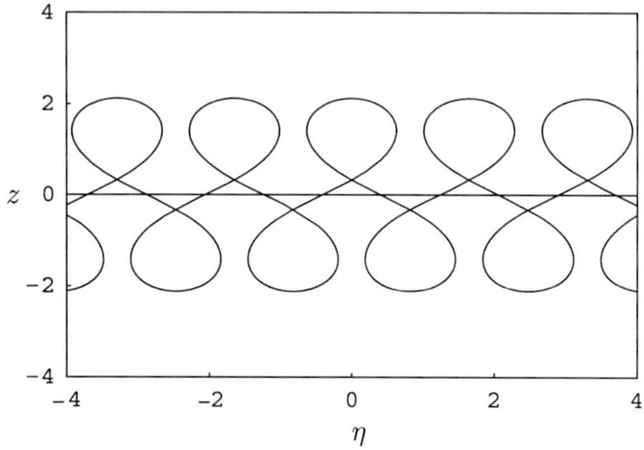

Figure 4.36: A typical periodic solution of the SPE with $c = -1$ comprising alternating upright loops and inverted loops for which $B = 2.5$ so that $m = 0.9$ and $\lambda = 1.64817$.

where

$$w = p\zeta, \quad p = \tfrac{1}{2}\sqrt{(z_4 - z_2)(z_3 - z_1)} \quad \text{and} \quad m = \frac{(z_3 - z_2)(z_4 - z_1)}{(z_4 - z_2)(z_3 - z_1)}. \tag{A.6}$$

In (A.5), $\mathrm{sn}(w|m)$ is a Jacobian elliptic function, where the notation is as used in Chapter 16 of [110]. On using result 400.01 in [44] we find from (A.5) and (A.3) that

$$\eta = \frac{1}{\varepsilon p}\,[wz_1 + (z_2 - z_1)\Pi(n; w|m)], \tag{A.7}$$

where $\Pi(n; w|m)$ is the incomplete elliptic integral of the third kind and the notation is as used in §17.2.15 of [110]. The solution to (A.1) is given in parametric form by (A.5) and (A.7) with w as the parameter. With respect to w, z in (A.5) is periodic with period $2K(m)$, where $K(m)$ is the complete elliptic integral of the first kind. It follows from (A.7) that the wavelength λ of the solution to (A.1) is

$$\lambda = \frac{2}{\varepsilon p}\left| z_1 K(m) + (z_2 - z_1)\Pi(n|m)\right|, \tag{A.8}$$

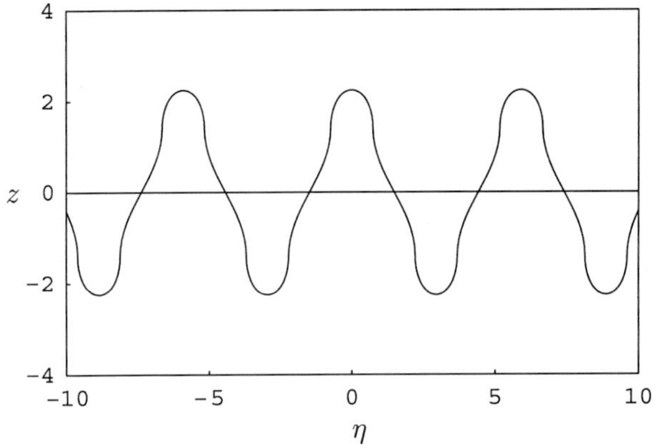

Figure 4.37: The composite periodic solution (4.6.18) of the SPE with $c = -1$ and $m = m_c$ so that $\lambda = 5.90747$.

where $\Pi(n|m)$ is the complete elliptic integral of the third kind. When $z_3 = z_4$, $m = 1$ and so (A.5) and (A.7) become

$$z = \frac{z_2 - z_1 n \tanh^2 w}{1 - n \tanh^2 w}, \quad \eta = \frac{1}{\varepsilon}\left[\frac{wz_3}{p} - 2\tanh^{-1}(\sqrt{n}\tanh w)\right]. \quad \text{(A.9)}$$

In (A.9), η was obtained by using

$$\Pi(n; w|1) = \frac{1}{1-n}\left[w - \sqrt{n}\tanh^{-1}(\sqrt{n}\tanh w)\right], \quad \text{(A.10)}$$

cf. result 111.04 in [44].

The second form of solution of (A.4) is found using result 255.00 in [44]. It is

$$z = \frac{z_3 - z_4 n\,\mathrm{sn}^2(w|m)}{1 - n\,\mathrm{sn}^2(w|m)} \quad \text{with} \quad n = \frac{z_3 - z_2}{z_4 - z_2}, \quad \text{(A.11)}$$

where w, p and m are as in (A.6). On using result 400.01 in [44] we find from (A.11) and (A.3) that

$$\eta = \frac{1}{\varepsilon p}\left[wz_4 - (z_4 - z_3)\Pi(n; w|m)\right]. \quad \text{(A.12)}$$

The solution to (A.1) is given in parametric form by (A.11) and (A.12) with w as the parameter. The wavelength of this solution is

$$\lambda = \frac{2}{\varepsilon p}\left| z_4 K(m) - (z_4 - z_3)\Pi(n|m)\right|. \tag{A.13}$$

When $z_1 = z_2$, $m = 1$ and so (A.11) and (A.12) become

$$z = \frac{z_3 - z_4 n \tanh^2 w}{1 - n\tanh^2 w}, \quad \eta = \frac{1}{\varepsilon}\left[\frac{w z_2}{p} + 2\tanh^{-1}(\sqrt{n}\tanh w)\right]. \tag{A.14}$$

$z_1 < z_2 < z_3 < z_4$	$z_1 < z_2 < z_3 = z_4$	$z_1 = z_2 < z_3 < z_4$	$z_1 = z_2 < z_3 = z_4$
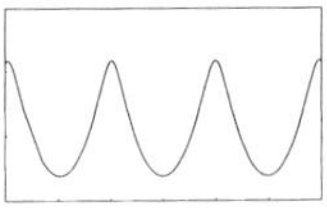 1.1 Periodic-hump solution $z_3 < 0$	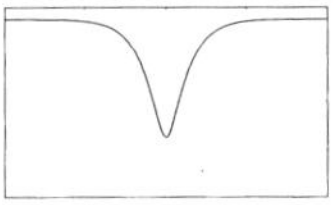 2.1 Inverted solitary smooth hump $z_3 = z_4 < 0$	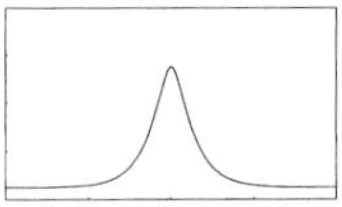 3.1 Hump-like solitary wave $z_3 < 0$	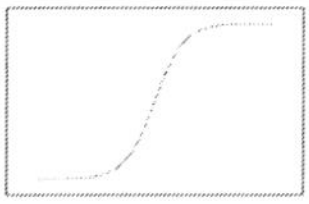 4.1 Kink-like solitary wave $z_3 = z_4 < 0$
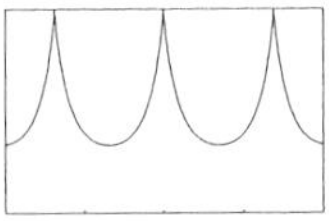 1.2 Periodic cuspon $z_3 = 0$	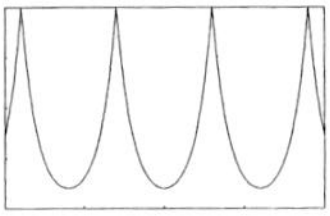 2.2 Periodic peakon $z_3 = z_4 = 0$	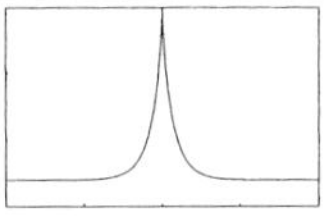 3.2 Solitary cuspon $z_3 = 0$	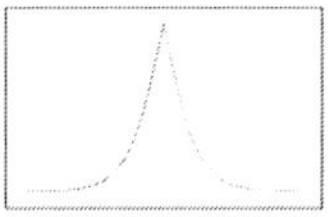 4.2 Singe peakon solution $z_3 = z_4 = 0$
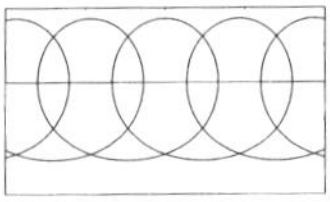 1.3a Periodic-loop solution $z_2 < 0, \quad z_3 > 0$		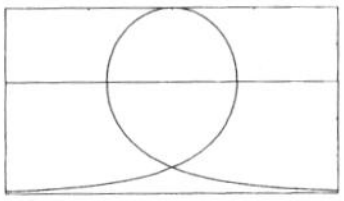 3.3 Loop-like solitary wave $z_2 < 0, \quad z_3 > 0$	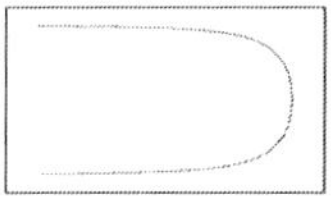 4.3a Two-valued solution $z_1 = z_2 < 0, \quad z_3 = z_4 > 0$
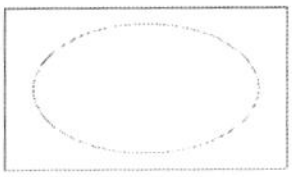 1.3b Closed curve $z_2 < 0, \quad z_3 > 0$			
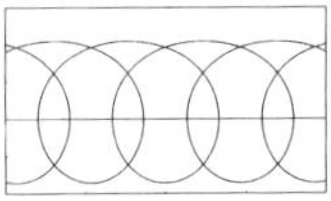 1.3c Periodic inverted loop solution $z_2 < 0, \quad z_3 > 0$	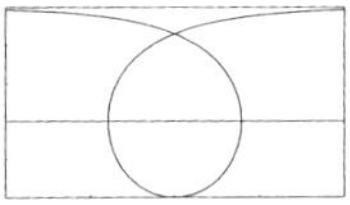 2.3 Inverted loop-like solitary wave $z_2 < 0, \quad z_3 = z_4 > 0$		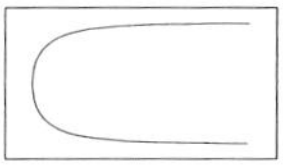 4.3b Two-valued solution $z_1 = z_2 < 0, \quad z_3 = z_4 > 0$
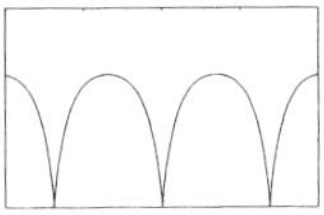 1.4 Periodic inverted cuspon $z_2 = 0$	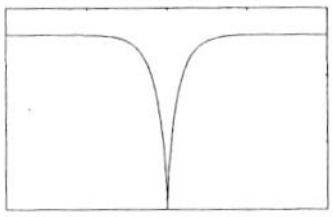 2.4 Solitary inverted cuspon $z_2 = 0, \quad z_3 = z_4 > 0$	Restriction $z_1 + z_2 + z_3 + z_4 = -2c$ can not be satisfied 3.4	Restriction $z_1 + z_2 + z_3 + z_4 = -2c$ can not be satisfied 4.4
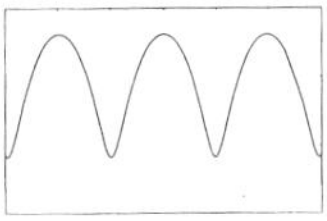 1.5 Periodic-hump solution $z_1 < 0, \quad z_2 > 0$	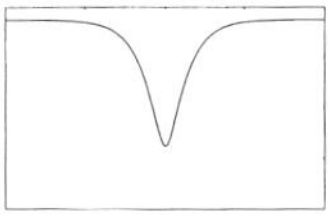 2.5 Inverted hump solitary wave $z_1 < 0, z_2 > 0, z_3 = z_4$	Restriction $z_1 + z_2 + z_3 + z_4 = -2c$ can not be satisfied 3.5	Restriction $z_1 + z_2 + z_3 + z_4 = -2c$ can not be satisfied 4.5

Table 4.1: Classification of travelling-wave solutions for the GDPE.

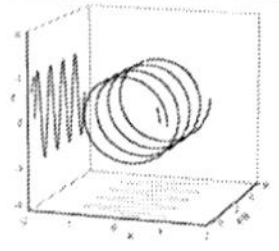 1.1 Periodic-hump solution	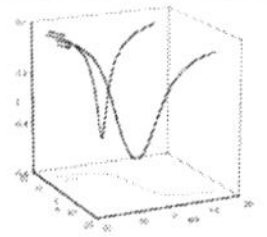 2.1 Inverted solitary smooth hump	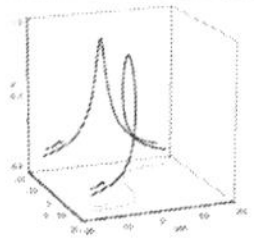 3.1 Hump-like solitary wave	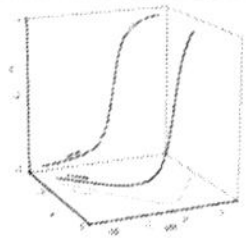 4.1 Kink-like solitary wave
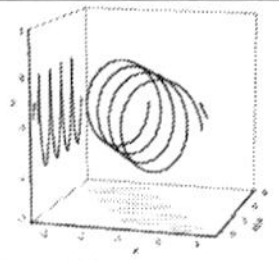 1.2 Periodic cuspon	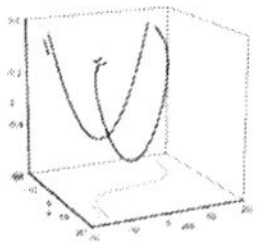 2.2 Periodic peakon	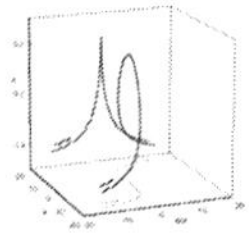 3.2 Solitary cuspon	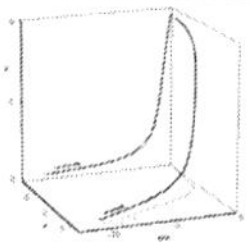 4.2 Singe peakon solution
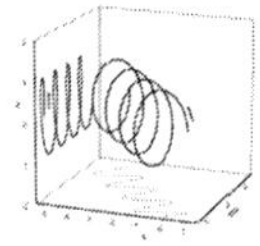 1.3a Periodic-loop solution		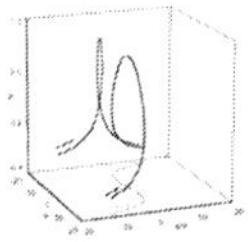 3.3 Loop-like solitary wave	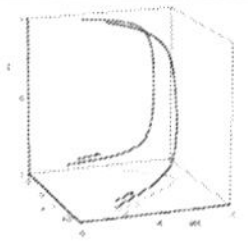 4.3a Two-valued solution
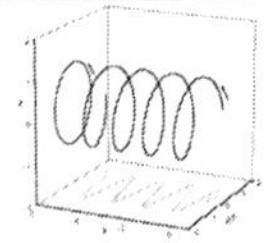 1.3b Closed curve			
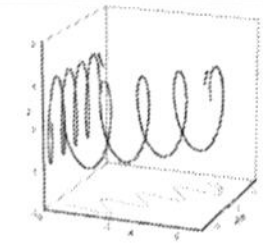 1.3c Periodic inverted loop solution	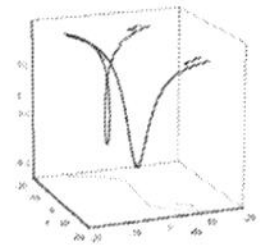 2.3 Inverted loop-like solitary wave		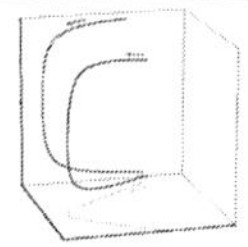 4.3b Two-valued solution
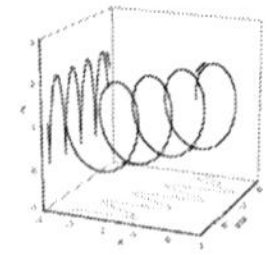 1.4 Periodic inverted cuspon	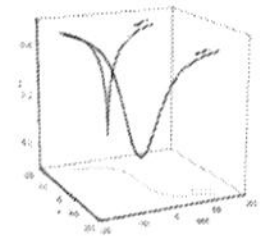 2.4 Solitary inverted cuspon	Restriction $z_1+z_2+z_3+z_4=-2c$ can not be satisfied 3.4	Restriction $z_1+z_2+z_3+z_4=-2c$ can not be satisfied 4.4
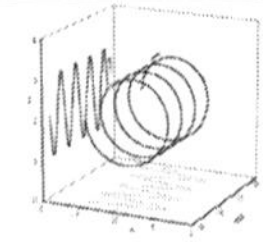 1.5 Periodic-hump solution	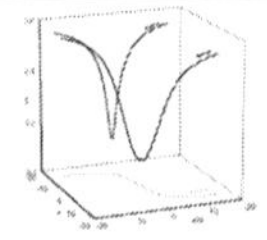 2.5 Inverted hump solitary wave	Restriction $z_1+z_2+z_3+z_4=-2c$ can not be satisfied 3.5	Restriction $z_1+z_2+z_3+z_4=-2c$ can not be satisfied 4.5

Table 4.2: Graphical interpretation of solutions for the GDPE.

Chapter 5
The Vakhnenko-Parkes equation

5.1 New independent coordinates

The multi-valued solutions obtained in Section 3.2 mean that the study of the VE (2.3.2), namely

$$\frac{\partial}{\partial x}\left(\frac{\partial}{\partial t}+u\frac{\partial}{\partial x}\right)u+u=0, \tag{5.1.1}$$

in the the original coordinates (x,t) leads to certain difficulties. These difficulties can be avoided by writing down the VE in new independent coordinates. We have succeeded in finding these coordinates. Historically, working separately, we (Vyacheslav Vakhnenko in the Ukraine and John Parkes in the UK) independently suggested such independent coordinates in which the solutions become one-valued functions. It is instructive to present the two derivations here. In one derivation a physical approach, namely a transformation between Euler and Lagrange coordinates, was used whereas in the other derivation a pure mathematical approach was used.

Let us define new independent variables (X,T) by the transformation

$$\varphi\, dT = dx - u\, dt, \qquad X = t. \tag{5.1.2}$$

The function φ is to be obtained. It is important that the functions $x=\theta(X,T)$ and $u=U(X,T)$ turn out to be single-valued. In terms of the coordinates (X,T) the solution of the VE (5.1.1) is given by single-valued parametric relations. The transformation into these coordinates is the key point in solving the problem of the interaction of solitons as well as explaining the multiple-valued solutions [11]. The transformation (5.1.2) is similar to the transformation between Eulerian coordinates (x, t) and Lagrangian coordinates (X, T). We require that $T = x$ if there is no perturbation, i.e. if $u(x, t) \equiv 0$. Hence $\varphi = 1$ when $u(x, t) \equiv 0$.

The function φ is the additional dependent variable in the equation system (5.1.4), (5.1.6) to which we reduce the original Eq. (5.1.1). We note that the transformation inverse to (5.1.2) is

$$dx = \varphi\, dT + U\, dX, \quad t = X, \quad U(X, T) \equiv u(x, t). \tag{5.1.3}$$

It follows that

$$\frac{\partial x}{\partial X} = U, \quad \frac{\partial x}{\partial T} = \varphi, \quad \frac{\partial t}{\partial X} = 1, \quad \frac{\partial t}{\partial T} = 0.$$

Hence

$$\frac{\partial \varphi}{\partial X} = \frac{\partial U}{\partial T} \tag{5.1.4}$$

and

$$\frac{\partial}{\partial X} = \frac{\partial}{\partial t} + u\frac{\partial}{\partial x}, \quad \frac{\partial}{\partial T} = \varphi\frac{\partial}{\partial x}. \tag{5.1.5}$$

By using (5.1.5), we can write Eq. (5.1.1) in terms of $\varphi(X, T)$ and $U(X, T)$, namely

$$U_{XT} + \varphi U = 0. \tag{5.1.6}$$

For use in Section 5.3, we note that, with (5.1.4), (5.1.6) may be written in the form

$$\varphi_{XX} + U\varphi = 0. \tag{5.1.7}$$

(5.1.4) and (5.1.6) or (5.1.7) is the main system of equations. It can be reduced to a nonlinear equation (5.1.10) in one unknown W defined by

$$W_X = U. \tag{5.1.8}$$

From (5.1.4) and (5.1.8) and the requirement that $\varphi = 1$ when $U \equiv 0$, we have

$$\varphi = 1 + W_T. \tag{5.1.9}$$

Then, by eliminating φ and U between (5.1.6), (5.1.8) and (5.1.9), we arrive at a transformed form of the VE (5.1.1), namely

$$W_{XXT} + (1 + W_T)W_X = 0. \tag{5.1.10}$$

Alternatively, by eliminating φ between (5.1.4) and (5.1.6), we obtain

$$UU_{XXT} - U_X U_{XT} + U^2 U_T = 0. \tag{5.1.11}$$

Furthermore it follows from (5.1.3) that the original independent coordinates (x, t) are given by

$$x = \theta(X, T) = x_0 + T + W, \qquad t = X, \tag{5.1.12}$$

where x_0 is an arbitrary constant. Since the functions $\theta(X, T)$ and $U(X, T)$ are single-valued, the problem of multi-valued solutions has been resolved from the mathematical point of view.

Alternatively, in a pure mathematical approach, we may start by introducing new independent variables X, T defined by

$$x = T + \int_{-\infty}^{X} U(X', T)\, dX' + x_1, \qquad t = X, \tag{5.1.13}$$

where x_1 is an arbitrary constant. From (5.1.13), we obtain (5.1.5) but with

$$\varphi(X, T) = 1 + \int_{-\infty}^{X} U_T\, dX'. \tag{5.1.14}$$

Now, on introducing (5.1.8), (5.1.13) and (5.1.14) may be identified with (5.1.12) and (5.1.9), respectively. The derivation of (5.1.10) and (5.1.11) proceeds as before.

The transformation into new coordinates, as has already been pointed out, was obtained by us independently of each other; nevertheless, we published the result together [74, 75, 76]. Following the papers [77, 78, 79, 80, 81, 82, 83], hereafter equation (5.1.10) (or in alternative form (5.1.11)) is referred to as the Vakhnenko-Parkes equation (VPE).

5.2 The travelling-wave solutions in new coordinates

In this Section we show that the travelling-wave solution (3.2.10) and (3.2.11) for equation (5.1.1) is also a travelling-wave solution when written in terms of the transformed coordinates (X,T). In order to do this, we need to express the independent variable ζ, as introduced in (3.2.8), in terms of X and T.

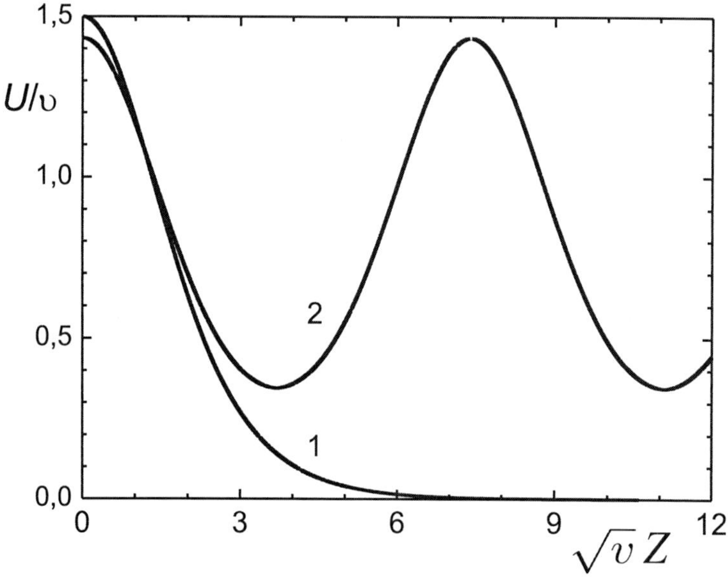

Figure 5.1: Travelling-wave solutions with $v > 0$ in coordinates (X, T).

From the expressions for z in (3.2.1) and (3.2.8), we obtain

$$\frac{d\eta}{d\zeta} = \frac{U - v}{|v|} \tag{5.2.1}$$

so that

$$|v|\eta = \int U\, d\zeta - v\zeta. \tag{5.2.2}$$

From the definition of η in (3.2.1), and the expressions for x and t

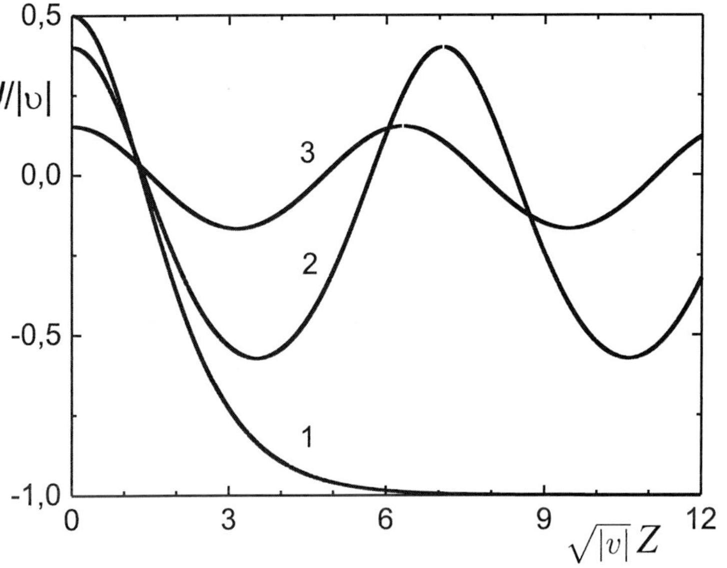

Figure 5.2: Travelling-wave solutions with $v < 0$ in coordinates (X, T).

given by (5.1.12), we obtain

$$|v|\eta = |v|^{1/2}[W - v(X - VT)], \text{ where } V := v^{-1}. \tag{5.2.3}$$

The expressions for $|v|\eta$ in (5.2.2) and (5.2.3) are equivalent if

$$\zeta = |v|^{1/2}Z, \text{ where } Z := X - VT - X_0 \tag{5.2.4}$$

and X_0 is an arbitrary constant, so that

$$W = \int U\,dZ \quad \text{and} \quad U = W_Z. \tag{5.2.5}$$

Hence, from (3.2.11) and (5.2.3), it follows that

$$W = \frac{\sqrt{|v|}}{p}\,[(z_1 + c)w + (z_3 - z_1)E(w|m)] + W_0, \quad \text{where} \quad w = p\sqrt{|v|}Z \tag{5.2.6}$$

and W_0 is an arbitrary constant. From (3.2.1) and (3.2.10), it follows that

$$\frac{U}{|v|} = c + z_3 - (z_3 - z_2)\,\mathrm{sn}^2(w|m), \quad \text{where} \quad w = p\sqrt{|v|}Z. \tag{5.2.7}$$

(5.2.6) and (5.2.7) give the travelling-wave solutions to the VPE in the forms (5.1.10) and (5.1.11), respectively. (5.2.7) is also the travelling-wave solution of the VE (5.1.1) expressed in terms of the new coordinates (X,T). In the limiting case $m = 1$, (5.2.7) gives a solitary wave in the following two forms: For $v > 0$

$$U/v = \tfrac{3}{2}\,\mathrm{sech}^2\left(\tfrac{1}{2}\sqrt{v}Z\right) \tag{5.2.8}$$

and, for $v < 0$,

$$U/|v| = -1 + \tfrac{3}{2}\,\mathrm{sech}^2\left(\tfrac{1}{2}\sqrt{|v|}Z\right). \tag{5.2.9}$$

These two solutions are illustrated by the curve 1 in Fig. 5.1 and Fig. 5.2, respectively. The other curves illustrate examples of the solution given by (5.2.7) when $m \neq 1$. The curves 1 and 2 in Fig. 5.1 relate to the curves 1 and 2, respectively, in Fig. 3.2. The curves 1, 2 and 3 in Fig. 5.2 relate to the curves 1, 2 and 3, respectively in Fig. 3.3.

There are two important observations to be made. Firstly, all the travelling-wave solutions in terms of the new coordinates are single-valued. Secondly, the periodic solution shown by curve 1 in Fig. 3.3, i.e. the solution consisting of parabolas, is not periodic in terms of the new coordinates. Hence, we reveal some accordance between curve 1 in Fig. 3.2 and curve 1 in Fig. 3.3. These features are important for finding the solutions by the inverse scattering method [76, 84, 85, 86, 87, 88, 89, 90] (see also Chapters 7–9).

5.3 The Vakhnenko-Parkes equation from the viewpoint of the inverse scattering method for the KdV equation

In this Section we use elements of the inverse scattering transform (IST) method as developed for the KdV equation [76]. The formulation of the IST method is discussed for the Vakhnenko-Parkes equation (VPE) in the form (5.1.10). It is shown that the equation system for the inverse scattering problem associated with the VPE cannot contain the isospectral Schrödinger equation. The results of this Section were completed before we made appreciable progress in formulation of the IST problem for the VPE.

As we will prove later in Section 7.1, the spectral problem associated with the VPE is of third order [84, 91, 92, 93]. At first reading, the present Section can be omitted. Nevertheless, methods stated here may be useful in the investigation of a new equation for which the spectral problem is unknown.

5.3.1 One-soliton solutions as reflectionless potentials

As was noted previously, the VE (see (2.3.2) or (5.1.1)) and the KdV equation

$$u_t + 6uu_x + u_{xxx} = 0 \tag{5.3.1}$$

have the same hydrodynamic nonlinearity and do not contain dissipative terms; only the dispersive terms are different. The similarity between these equations indicates that, in studying the VE (5.1.1) and the VPE (5.1.10), the application of the IST method should be possible. The IST method is the most appropriate way of tackling initial value problems. The results of applying the IST method would be useful in solving the Cauchy problem for both the VE and the VPE. The study of the VPE is of scientific interest from the viewpoint of the general problem of integrability of nonlinear equations.

The method of the IST is a powerful method to use as a means for solving nonlinear differential equations. Let us recall that the KdV equation (5.3.1) is associated with the system of the equations (the Lax pair)

$$\psi_{xx} + u\psi = \lambda\psi, \tag{5.3.2}$$

$$\psi_t + 3\lambda\psi_x + \psi_{xxx} + 3u\psi_x = 0. \tag{5.3.3}$$

The equation system (5.3.2), (5.3.3) is a case of the IST method presented in the classic paper [34] by Gardner et al. Since the system (5.3.2), (5.3.3) contains the Schrödinger equation (5.3.2), we will use the elements of the IST method as applied to the KdV equation in order to analyze the VPE. The known one-soliton solution of the KdV equation (5.3.1) has the form (without the time-dependence)

$$u = 2\varrho^2 \operatorname{sech}^2 \varrho x. \tag{5.3.4}$$

Here, as an example, we will consider the case $\varrho = 1$.

The results in Section 5.3 are based on the assumption that the system of equations associated with the VPE (5.1.10), which are analogous to (5.3.2) and (5.3.3), are unknown.

Now let us focus on the fact that Eq. (5.1.7) is the Schrödinger equation

$$\frac{\partial^2 \psi}{\partial X^2} - Q\psi = \lambda\psi$$

with the eigenvalue (energy) $\lambda = 0$ and potential $Q = -U$. Equation (5.1.7) determines the dependence on the coordinate X, and time T appears here as a parameter. However, the time-dependence is determined by Eq. (5.1.4).

The known one-soliton solution of Eq. (5.1.10), namely (5.2.8) which we obtained in Section 5.2, has the form

$$U = \frac{3v}{2} \operatorname{sech}^2\left(\frac{vX - T}{2\sqrt{v}}\right). \tag{5.3.5}$$

If it is not otherwise noted, for convenience here we will consider $v = 4$, $T = 0$, and then Eq. (5.3.5) reduces to

$$U = 6 \operatorname{sech}^2 X. \tag{5.3.6}$$

The principal fact is that both $u = 2 \operatorname{sech}^2 x$ from (5.3.4) and $U = 6 \operatorname{sech}^2 X$ from (5.3.6) relate to reflectionless potentials. The general form of the reflectionless potentials is (see Section 2.4 in [27])

$$u = m(m+1) \operatorname{sech}^2 x. \tag{5.3.7}$$

We have $m = 1$ for the potential (5.3.4) and $m = 2$ for the potential (5.3.6). It is known [27, 28] that for integrable nonlinear equations, reflectionless potentials generate soliton solutions (which, in the general case, are N-soliton solutions).

5.3.2 Two-level reflectionless potential

Let us consider the one-soliton solution of the system (5.3.2), (5.3.3) in the framework of the IST method for the KdV equation. For this purpose let us analyze the Schrödinger equation with the potential $Q \equiv -U = -6\,\text{sech}^2 X$, namely

$$\frac{d^2\psi}{dX^2} - Q\psi = -k^2\psi, \qquad k^2 = -\lambda. \tag{5.3.8}$$

(Recall that here T is a parameter.) For the scattering problem, the solution of Eq. (5.3.8) should satisfy the boundary conditions

$$\psi(X,k) = \begin{cases} e^{-ikX}, & X \to -\infty \\ b(k)\,e^{ikX} + a(k)\,e^{-ikX}, & X \to +\infty \end{cases}, \tag{5.3.9}$$

where $b(k)$ and $a(k)$ are the coefficients of reflection and transmission, respectively.

In Section 2.4 in [27], the original method for finding the wave-functions ψ and eigenvalues for the reflectionless potential $Q_m = -m(m+1)\,\text{sech}^2 X$ was described. The general solution y_m of Eq. (5.3.8) for the potential Q_m connects with the general solution Y_0 for $Q_0 = 0$ by the relationship

$$y_m(X,k) = \prod_{m'=1}^{m} \left(m' \tanh X - \frac{d}{dX} \right) Y_0(X,k), \tag{5.3.10}$$

and then

$$a(k) = \prod_{m'=1}^{m} \frac{ik + m'}{ik - m'}, \qquad b(k) = 0. \tag{5.3.11}$$

In our case ($m = 2$), Eq. (5.3.8) has two bound states, namely

$$\begin{aligned} -ik_1 \equiv \kappa_1 = 1, \qquad & \psi_1 = \sqrt{\tfrac{3}{2}}\,\tanh X\,\text{sech}\,X, \\ -ik_2 \equiv \kappa_2 = 2, \qquad & \psi_2 = \tfrac{\sqrt{3}}{2}\,\text{sech}^2 X. \end{aligned} \tag{5.3.12}$$

The wave-functions ψ_i are normalized, i.e. $\int_{-\infty}^{+\infty} |\psi_i|^2 dX = 1$, and this conforms to the requirement used in the IST method.

Here the main difference between the VPE and known integrable nonlinear equations appears. It is connected with the existence of only one bound state for the known equations associated with the isospectral Schrödinger equation, while for the VPE two bound states occur. Indeed, for the known integrable equations, the potential corresponding to the one-soliton solution has the following dependence on the space coordinate (see Eq. (4.3.9) in [27])

$$u(x) = 2\varrho^2 \operatorname{sech}^2 \varrho x. \tag{5.3.13}$$

It is easy to see that this is related to the case $m = 1$ in Eq. (5.3.7), i.e. there is only the one bound state

$$\psi = \sqrt{\varrho/2}\, \operatorname{sech}\, \varrho x, \tag{5.3.14}$$

$$\psi \to c\sqrt{\varrho}\exp(-\varrho x), \quad c = \sqrt{2}, \quad \text{as} \quad x \to +\infty. \tag{5.3.15}$$

5.3.3 Reconstruction of the one-soliton solution for the VPE

Keeping in mind that there is an incomplete analogy of our problem to known integrable equations, we shall try to reconstruct the potential (the solution of the VPE) from the scattering data as well as to find afterwards the time-dependence for the scattering data and for the one-soliton solution.

As is well known [27, 28], in order to reconstruct the potential for the Schrödinger equation (5.3.8), we have to know the scattering data. From the relationships (5.3.12) we obtain, as $X \to \infty$,

$$\begin{aligned} &\psi_1 \to c_1 e^{-\kappa_1 X}, \qquad c_1 = \sqrt{6}, \qquad \kappa_1 = 1, \\ &\psi_2 \to c_2 e^{-\kappa_2 X}, \qquad c_2 = \sqrt{12}, \quad\ \kappa_2 = 2. \end{aligned} \tag{5.3.16}$$

Clearly, $\kappa_1 = \frac{1}{2}\kappa_2 = 1$ is in agreement with (5.3.5), (5.3.6) and (5.3.4). However, we shall abandon this condition, i.e. $v = 4$ in Eq. (5.3.5), and in the final formulas.

For convenience we reproduce the well known procedure for the reconstruction of the potential. The function $B(X;T)$ is constructed from the scattering data (T is the parameter)

$$B(X;T) = \sum_{m=1}^{n} c_m^2(t) e^{-\kappa_m X} + \frac{1}{2\pi} \int_{-\infty}^{+\infty} b(k,T) e^{ikX}\, dk.$$

In the next step, the following Marchenko–Gelfand–Levitan equation is to be solved [95] for the unknown $K(X, y; T)$:

$$K(X,y;T)+B(X+y;T)+\int_X^{+\infty} B(y+z;T)K(X,z;T)\,dz = 0. \quad (5.3.17)$$

The potential is then obtained by means of the relationship

$$-U = Q = -2\frac{d}{dX}K(X,X;T). \quad (5.3.18)$$

In particular, for the reflectionless potential (5.3.7), $b(k) = 0$ in (5.3.9), and the solution can be found in the form

$$K(X,y;T) = -\sum_{m=1}^{N} c_m(T)\psi_m(X;T)e^{-\kappa_m y}. \quad (5.3.19)$$

This procedure, as is well known, leads to the equation system in ψ_m

$$\mathbf{A}\mathbf{\Psi} = \mathbf{C}, \quad (5.3.20)$$

where the matrix $\mathbf{A} = [a_{mn}]$ has elements

$$a_{mn} = \delta_{mn} + c_n(T)c_m(T)\,\frac{e^{-X(\kappa_m+\kappa_n)}}{\kappa_m + \kappa_n},$$

and $\mathbf{\Psi} = [\psi_m]$ and $\mathbf{C} = [c_m(T)e^{-\kappa_m X}]$ are column-vectors.

In Eqs. (5.3.17)–(5.3.20), T is a parameter. Although we took $T = 0$ earlier, we preserve the variable T in these relationships in order to use them later to find the time-dependence of the scattering data.

It is known [27, 28] that for a reflectionless potential the value of the determinant $\Delta = \det[a_{mn}]$ is sufficient for reconstructing the potential. Then Eq. (5.3.18) is reduced to

$$K(X,X;T) = \frac{d\ln|\Delta|}{dX}, \qquad -U = -2\frac{d^2\ln|\Delta|}{dX^2}. \quad (5.3.21)$$

We use (5.3.20), (5.3.21) to obtain the one-soliton solution of the VPE. The scattering data (5.3.16) and $b(k) = 0$ enable us to define the determinant

$$\Delta = \begin{vmatrix} 1+\frac{c_1^2}{2}e^{-2X} & \frac{c_1c_2}{3}e^{-3X} \\ \frac{c_1c_2}{3}e^{-3X} & 1+\frac{c_2^2}{4}e^{-4X} \end{vmatrix} = (1+e^{-2X})^3 \quad (5.3.22)$$

and then the potential

$$-U = 12\frac{d}{dX}\left(\frac{e^{-2X}}{1+e^{-2X}}\right) = -6\,\mathrm{sech}^2\,X.$$

Thus we have repeated the standard method for reproducing the potential by means of scattering data (as yet without time-dependence). It is clear from $U = W_X$ and (5.3.21) that

$$W = 2K(X, X; T).$$

It is noted that the determinant for the one-soliton solution of the KdV equation (5.3.1) has the form

$$\Delta = 1 + e^{-2x}, \quad u = 2\,\mathrm{sech}^2\,x. \tag{5.3.23}$$

The interpretation of (5.3.22) is important. In the matrix, two states (5.3.16) are involved. Clearly, the time-dependence for an individual state is its own characteristic. However, since these two states relate to the common soliton, there must be a connection between them, i.e. $c_1(T)$ and $c_2(T)$ must be connected. The relation (5.3.22) determines this connection.

In the first instance we considered the dependence of the potential on the space coordinate, and the time was a parameter. Let us now find the time-dependence of the scattering data $c_1(T)$, $c_2(T)$ that enables us to find the functional dependence of the potential (5.3.6) on T, i.e. the time-dependence of the one-soliton solution. We start from the relation (see Eq. (22), Chap. 1, Section 2 in [28])

$$\psi(X, k; T) = e^{-ikX} + \int_X^{+\infty} K(X, y; T)e^{-iky}dy. \tag{5.3.24}$$

Hence, there is a linear operator that reduces the solution e^{-ikX} of the Schrödinger equation with null potential $Q = 0$ to the solution of this equation with the potential $U(X)$. The function $K(X, y; T)$ is the kernel of the transformation operator.

We write Eq. (5.3.24) for $k = 0$; this procedure is correct and an appropriate theorem has been proved (see Section 3.3 in [27])

$$\psi(X, k = 0; T) = 1 + \int_X^{+\infty} K(X, y; T)dy. \tag{5.3.25}$$

Clearly, $\psi(X,k=0;T)=\varphi(X,T)$, where $\varphi(X,T)$ satisfies the equation system (5.1.4), (5.1.7). Taking into account (5.3.18) and (5.3.25), we obtain from the relationship (5.1.4)

$$1+\int_X^{+\infty} K(X,y;T)dy = 2\frac{\partial K(X,X,T)}{\partial T}+C. \tag{5.3.26}$$

Since this equation must be valid at arbitrary X, and taking into account that the function $K(X,y;T)\to 0$ at $|X|\to\infty$, we define the constant of integration $C=1$. We write, once again, $K(X,y;T)$ as in (5.3.19) because the potential is reflectionless, and from (5.3.26) we obtain

$$\sum_{m=1}^{2}\frac{c_m(T)}{\kappa_m}\psi_m(X;T)e^{-\kappa_m X}=2\sum_{m=1}^{2}\frac{\partial c_m(T)\psi_m(X;T)}{\partial T}e^{-\kappa_m X}. \tag{5.3.27}$$

In this equation we must substitute the values ψ_m that are the solution of system (5.3.20). Here we consider the values c_m already as functions of T, i.e. $c_m=c_m(T)$. For example ψ_1 is given by

$$\psi_1=\Delta^{-1}\left(c_1e^{-\kappa_1 X}+\frac{c_1c_2^2}{2\kappa_2}e^{-(\kappa_1+2\kappa_2)X}-\frac{c_1c_2^2}{\kappa_1+\kappa_2}e^{-(\kappa_1+2\kappa_2)X}\right). \tag{5.3.28}$$

Here Δ is the determinant (5.3.22) with time-dependence of $c_m=c_m(T)$. We can calculate the following terms which are required for (5.3.27) (with $\kappa_1=1,\ \kappa_2=2$):

$$\begin{aligned}&\sum_{m=1}^{2}\frac{c_m(T)}{\kappa_m}\psi_m(X;T)e^{-\kappa_m X}=\Delta^{-1}\left(c_1^2e^{-2X}+\tfrac{1}{2}c_2^2e^{-4X}\right),\\&\sum_{m=1}^{2}c_m(T)\psi_m(X;T)e^{-\kappa_m X}=\Delta^{-1}\left(c_1^2e^{-2X}+c_2^2e^{-4X}+\tfrac{1}{12}c_1^2c_2^2e^{-6X}\right).\end{aligned} \tag{5.3.29}$$

Then, on substituting (5.3.29) into (5.3.27) and equating to zero the coefficients of e^{-2jX}, $(j=1,\dots,6)$, we obtain the following system of differential equations for $c_m(T)$, $(m = 1, 2)$:

$$
\begin{aligned}
e^{-2X}: &\quad \left(c_1^2\right)' = \tfrac{1}{2}c_1^2, \\
e^{-4X}: &\quad \left(c_2^2\right)' = \tfrac{1}{4}(c_2^2 + c_1^4), \\
e^{-6X}: &\quad \tfrac{1}{3}(c_1^2c_2^2)' + c_1^2\left(c_2^2\right)' - c_2^2\left(c_1^2\right)' = c_1^2c_2^2, \\
e^{-8X}: &\quad c_1^2\left(c_1^2c_2^2\right)' - c_1^2c_2^2\left(c_1^2\right)' = \tfrac{1}{4}\left(c_1^4c_2^2 + 9c_2^4\right), \\
e^{-10X}: &\quad c_2^2\left(c_1^2c_2^2\right)' - c_1^2c_2^2\left(c_2^2\right)' = \tfrac{1}{2}c_1^2c_2^4, \\
e^{-12X}: &\quad c_1^2c_2^2\left(c_1^2c_2^2\right)' = c_1^2c_2^2\left(c_1^2c_2^2\right)',
\end{aligned}
\tag{5.3.30}
$$

where the prime denotes the derivative with respect to time T.

The equation system (5.3.30) is an over-determined one; only the first two equations are independent. Consequently, we solve them with initial conditions $c_1^2(0) = 6$, $c_2^2(0) = 12$. At first, we write the general solution of the system (5.3.30) as

$$c_1^2(T) = r_1e^{T/2}, \quad c_2^2(T) = r_2e^{T/4} + \tfrac{1}{3}r_1^2e^T, \tag{5.3.31}$$

where r_1, r_2 are arbitrary constants. Hence, in the general case, the time-dependence of the first and second states are different. Nevertheless, we have $r_2 \equiv 0$ due to the relationship between $c_1(0)$ and $c_2(0)$ and then

$$c_1^2(T) = c_1^2(0)e^{T/2} = 6e^{T/2}, \quad c_2^2(T) = \tfrac{1}{3}c_1^4(0)e^T = 12e^T. \tag{5.3.32}$$

Thus, the time-dependences satisfy the condition $c_1^2(T)/c_2(T) = \text{const}$. Indeed, if the time-dependences is as in (5.3.32), the determinant (5.3.22) can be rewritten as a perfect cube, namely

$$\Delta = \left(1 + e^{-2(X-T/4)}\right)^3. \tag{5.3.33}$$

For convenience, up to this point we have used $\kappa_1 = 1$, $\kappa_2 = 2$. Now we return to one arbitrary parameter κ_1 (with $\kappa_2 = 2\kappa_1$) and rename it as $\alpha \equiv \kappa_1$, and then we obtain

$$\Delta = \left\{1 + \exp\left[-2\alpha\left(X - \frac{T}{4\alpha^2}\right)\right]\right\}^3. \tag{5.3.34}$$

The potential for the one-soliton solution can easily be found by use of Eq. (5.3.21) so that

$$U = 2\frac{d^2\ln|\Delta|}{dX^2} = 6\alpha^2\operatorname{sech}^2\Theta, \qquad \Theta = \alpha\left(X - X_0 - \frac{T}{4\alpha^2}\right). \tag{5.3.35}$$

This is the one-soliton solution for the VPE.

For reference we give the complete equations for finding the solution of the VE (2.3.2) in terms of the initial variables x, t, (for convenience T is renamed as $\mu \equiv T$ because here μ is a parameter):

$$\frac{\partial}{\partial x}\left(\frac{\partial}{\partial t}+u\frac{\partial}{\partial x}\right)u+u=0,$$

$$u=\left(\frac{\partial W}{\partial t}\right)_{\mu}, \quad x=x_0+\mu+W, \quad W=2\left(\frac{\partial \ln|\Delta|}{\partial t}\right)_{\mu}, \tag{5.3.36}$$

$$\Delta=(1+q^2)^3, \quad q=\exp(-\Theta), \quad \Theta=\alpha\left(t-\frac{\mu-\mu_0}{4\alpha^2}\right), \tag{5.3.37}$$

$$\alpha=\text{const.}, \quad \mu_0=\text{const.}$$

Thus we have obtained the one-soliton solution of the VE as well as the VPE using elements of the IST method for the KdV equation. The proposed method is also applicable for finding the two-soliton solution. It is likely that this procedure will shed light upon the formulation of the IST problem that enable one to make progress in the study of the Cauchy problem for the VE (2.3.2).

5.3.4 Two-soliton solution

Let us consider the two-soliton solution for the VE (2.3.2). The key for constructing this solution is the value which is assigned to the determinant in (5.3.36) in the one-soliton solution. For information we rewrite the values (5.3.37) once again

$$\Delta=(1+q^2)^3, \quad q=\exp\left[-\alpha\left(t-\frac{\mu-\mu_0}{4\alpha^2}\right)\right]. \tag{5.3.38}$$

It can be seen that there is the some analogy to the one-soliton solution of the KdV equation (5.3.23), namely

$$\Delta=1+q^2, \quad q=\exp(\alpha x-4\alpha^3 t).$$

Moreover, as we noted, the potentials corresponding to the one-soliton solution

(a) for the VPE ($T = 0,\ \alpha = 1$)

$$U = 6\,\mathrm{sech}^2 X, \tag{5.3.39}$$

(b) for the KdV equation ($t = 0,\ \varrho = 1$)

$$u = 2\,\mathrm{sech}^2 x, \tag{5.3.40}$$

differ from each other by their coefficients. Bearing in mind Eq. (5.3.18) and that $K = (\ln|\Delta|)_X$ (see (5.3.21)), one can see that the coefficient 6 in (5.3.39), in contrast to the coefficient 2 in (5.3.40), is generated by the exponent 3 in the relationship (5.3.38).

Now, if it is recalled that the two-soliton solution for the KdV has the form [36]

$$\begin{aligned} \tilde{F} &= \Delta = 1 + q_1^2 + q_2^2 + \tilde{A}_{12} q_1^2 q_2^2, \\ \tilde{A}_{12} &= \frac{(\alpha_1 - \alpha_2)^2}{(\alpha_1 + \alpha_2)^2}, \\ q_i &= \exp[\alpha_i (x - x_{0i}) - 4\alpha_i^3 t], \end{aligned} \tag{5.3.41}$$

we can expect that the two-soliton solution for the VE can be found in the form (5.3.36) with the following value of F instead of Δ in relation (5.3.21):

$$F = \left(1 + q_1^2 + q_2^2 + A_{12} q_1^2 q_2^2\right)^3, \quad q_i = \exp\left[-\alpha_i \left(t - \frac{\mu - \mu_i}{4\alpha_i^2}\right)\right]. \tag{5.3.42}$$

The value of A_{12} is to be determined. It should be noted that F is not equal to the determinant Δ of the matrix in (5.3.20) which is constructed from four states with q_1, q_1^2, q_2, q_2^2, (each soliton has two bound states (5.3.12)), namely

$$\Delta = \begin{vmatrix} 1 + 3q_1^2 & 2\sqrt{2} q_1^3 & \frac{6\sqrt{\alpha_1\alpha_2}}{\alpha_1 + \alpha_2} q_1 q_2 & \frac{6\sqrt{2\alpha_1\alpha_2}}{\alpha_1 + 2\alpha_2} q_1 q_2^2 \\ 2\sqrt{2} q_1^3 & 1 + 3q_1^4 & \frac{6\sqrt{2\alpha_1\alpha_2}}{2\alpha_1 + \alpha_2} q_1^2 q_2 & \frac{6\sqrt{\alpha_1\alpha_2}}{\alpha_1 + \alpha_2} q_1^2 q_2^2 \\ \frac{6\sqrt{\alpha_1\alpha_2}}{\alpha_1 + \alpha_2} q_1 q_2 & \frac{6\sqrt{2\alpha_1\alpha_2}}{2\alpha_1 + \alpha_2} q_1^2 q_2 & 1 + 3q_2^2 & 2\sqrt{2} q_2^3 \\ \frac{6\sqrt{2\alpha_1\alpha_2}}{\alpha_1 + 2\alpha_2} q_1 q_2^2 & \frac{6\sqrt{\alpha_1\alpha_2}}{\alpha_1 + \alpha_2} q_1^2 q_2^2 & 2\sqrt{2} q_2^3 & 1 + 3q_2^4 \end{vmatrix}. \tag{5.3.43}$$

If the relation $F = \Delta$ were true, we would have $A_{12} = \tilde{A}_{12}$. Moreover, these conditions would lead us to the statement that the problem for the scattering data for the VE (2.3.2) should connect with the isospectral Schrödinger equation. This statement was made in the paper by Hirota and Satsuma [96] as well as in the monograph by Newell (see Chaps. 3 and 4 in [29]). However, because $F \neq \Delta$ and $A_{12} \neq \tilde{A}_{12}$, we can state that the equation system for the IST problem associated with the VPE (5.1.10) does not contain the isospectral Schrödinger equation.

The value A_{12} for (5.3.42) can be determined in the following way. The functional relation (5.3.42), with A_{12} regarded as unknown, is substituted into Eq. (5.3.36), and then into Eq. (5.1.10). Equating to zero the coefficients of $\exp[-2(i\alpha_1+j\alpha_2)X]$, $(i,j=0,\ldots,4,\ i+j\neq 0)$, we obtain a system of equations in one unknown A_{12}. It turns out that the equations are dependent. As a result we obtain

$$A_{12} = \frac{(\alpha_1-\alpha_2)^2}{(\alpha_1+\alpha_2)^2}\cdot\frac{\alpha_1^2+\alpha_2^2-\alpha_1\alpha_2}{\alpha_1^2+\alpha_2^2+\alpha_1\alpha_2}. \tag{5.3.44}$$

Thus the relationships (5.3.36) (with F instead of Δ), (5.3.42) and (5.3.44) are the exact two-soliton solution of the VE (2.3.2). In terms of x and t, we have

$$\frac{\partial}{\partial x}\left(\frac{\partial}{\partial t}+u\frac{\partial}{\partial x}\right)u+u=0,$$

$$u=\left(\frac{\partial W}{\partial t}\right)_\mu,\quad x=x_0+\mu+W,\quad W=2\left(\frac{\partial \ln|F|}{\partial t}\right)_\mu, \tag{5.3.45}$$

$$F=\left(1+q_1^2+q_2^2+A_{12}q_1^2q_2^2\right)^3,\quad q_i=\exp(-\Theta_i), \tag{5.3.46}$$

$$A_{12} = \frac{(\alpha_1-\alpha_2)^2}{(\alpha_1+\alpha_2)^2}\cdot\frac{\alpha_1^2+\alpha_2^2-\alpha_1\alpha_2}{\alpha_1^2+\alpha_2^2+\alpha_1\alpha_2},$$

$$\Theta_i=\alpha_i t-\frac{\mu-\mu_i}{4\alpha_i},\quad \alpha_i=\text{const.},\quad \mu_i=\text{const.}$$

The function $W(X,T)$ in space (X,T) is the two-soliton solution for the VPE.

An equivalent result has been obtained, independently of the method presented here, in Section 6.5.4 by the means of the Hirota method [96, 35, 36] in terms of other variables.

5.3.5 Concluding remarks

The main result of Section 5.3 is that we have obtained a way of applying the IST method to the VPE. Keeping in mind that the IST is the most appropriate way of tackling the initial value problem, one has to formulate the associated eigenvalue problem. We have proved that the equation system for the IST problem associated with the VPE does not contain the isospectral Schrödinger equation. Nevertheless, the analysis of the VPE in the context of the isospectral Schrödinger equation allowed us to obtain the two-soliton solution. Thus the results stated here may be useful in the investigation of a new equation for which the spectral problem is unknown.

Historically, once this investigation was completed, we were able to make some progress in the formulation of the IST for the VPE. In Section 7.1 we will prove that the spectral problem associated with the VPE is of third order.

Chapter 6
The Hirota method

Now let us define the notion 'soliton' more precisely. Apart from the fact that a soliton is a stable solitary wave with particle-like properties, a soliton must possess additional properties. One property is that two such solitary waves may pass through each other without any loss of identity. Consider two solitons with different speeds, the faster one chasing the slower one. The faster soliton will eventually overtake the slower one. After the nonlinear interaction, two solitons again will emerge, with the faster one in front, and each will regain its former identity precisely. The only interaction memory will be a phase shift; each soliton will be centered at a location different from where it would have been had it traveled unimpeded. However, this property is still not sufficient in order that the solitary wave be a soliton. There are equations which possess solutions which are a nonlinear superposition of two solitary waves but which do not have all the properties enjoyed by soliton equations. A soliton equation, when it admits solitary wave solutions, must possess a solution which satisfies the 'N-soliton condition' (see Section 6.5.1). The solitary wave with these properties defines a soliton. The term 'soliton' was originally coined by Zabusky and Kruskal in 1965 [111].

One of the key properties of a soliton equation is that it has an infinite number of conservation laws. These soliton equations satisfy the Hirota condition ('N-soliton condition') and are exactly integrable.

The Hirota method not only gives the N-soliton solution, but enables one to find a way from the Bäcklund transformation through the conservation laws and associated eigenvalue problem to the inverse scattering method. Thus the Hirota method, which can be applied only for finding solitary wave solutions or traveling wave solutions, allows us to formulate the inverse scattering method which is the most appropriate way of tackling the initial value problem (Cauchy problem). Consequently, in this case, the integrability of an equation can be regarded as proved.

6.1 The D-operator and N-soliton solution

Various effective approaches have been developed to construct exact wave solutions of completely integrable equations. One of the fundamental direct methods is undoubtedly the Hirota bilinear method [35, 36, 96, 112], which possesses significant features that make it practical for the determination of multiple soliton solutions.

In the Hirota method, the equation under investigation should first be transformed, if possible, into Hirota bilinear form [35]. There are several possibilities, the most common of which is

$$F(D_X, D_T) f \cdot f = 0, \tag{6.1.1}$$

where F is a polynomial in D_T and D_X. Each equation has its own polynomial. The Hirota bilinear D-operator is defined as (see Section 5.2 in [35])

$$D_T^n D_X^m a \cdot b = \tag{6.1.2}$$

$$= \left(\frac{\partial}{\partial T} - \frac{\partial}{\partial T'}\right)^n \left(\frac{\partial}{\partial X} - \frac{\partial}{\partial X'}\right)^m a(T, X) b(T', X')\Big|_{T=T',\, X=X'}.$$

If the polynomial F satisfies conditions (see (5.41), (5.42) in [35])

$$F(D_X, D_T) = F(-D_X, -D_T), \quad F(0, 0) = 0, \tag{6.1.3}$$

then the Hirota method can be applied successfully.

According to [35], the N-soliton solution reads as follows

$$f = \sum_{\mu=0,1} \exp\left[2\left(\sum_{i=1}^{N} \mu_i \eta_i + \sum_{i<j}^{(N)} \mu_i \mu_j \ln b_{ij}\right)\right], \tag{6.1.4}$$

where

$$b_{ij}^2 = -\frac{F[2(k_i - k_j), -2(\omega_i - \omega_j)]}{F[2(k_i + k_j), -2(\omega_i + \omega_j)]}, \quad \eta_i = k_i X - \omega_i T + \alpha_i. \tag{6.1.5}$$

The connection between k_i and ω_i is found by the dispersion relations

$$F(2k_i, -2\omega_i) = 0, \quad i = 1, \ldots, N. \tag{6.1.6}$$

In (6.1.4), $\sum_{\mu=0,1}$ means the summation over all possible combinations of $\mu_1 = 0$ or 1, $\mu_2 = 0$ or $1, \ldots, \mu_N = 0$ or 1, and $\sum_{i<j}^{(N)}$ means the summation over all possible combinations of N elements under the condition $i < j$.

Moreover, for there to be an N-soliton solution (NSS) to (6.1.1) with $N(\geq 1)$ arbitrary, $F(D_X, D_T)$ must satisfy the 'N-soliton condition' (NSC) [35], namely

$$G^{(n)}(p_1, \ldots, p_n) = 0, \qquad n = 1, 2, \ldots, N, \tag{6.1.7}$$

where

$$G^{(1)}(p_1) := 0 \tag{6.1.8}$$

and, for $n \geq 2$,

$$\begin{aligned} G^{(n)}(p_1, \ldots, p_n) :&= C \textstyle\sum_{\sigma=\pm 1} \Big\{ F\Big(\sum_{i=1}^{n} \sigma_i p_i, \sum_{i=1}^{n} \sigma_i \Omega_i \Big) \\ &\quad \times \textstyle\prod_{i>j}^{(n)} F(\sigma_i p_i - \sigma_j p_j, \sigma_i \Omega_i - \sigma_j \Omega_j) \sigma_i \sigma_j \Big\}. \end{aligned} \tag{6.1.9}$$

In (6.1.9), the Ω_i are given in terms of the p_i by the dispersion relations $F(p_i, \Omega_i) = 0$ $(i = 1, \ldots, N)$, $\sum_{\sigma=\pm 1}$ means the summation over all possible combinations of $\sigma_1 = \pm 1$, $\sigma_2 = \pm 1, \ldots, \sigma_n = \pm 1$, and C is a function of the p_i that is independent of the summation indices $\sigma_1, \ldots, \sigma_n$.

From (6.1.8) it follows that (6.1.7) is satisfied for $n = 1$. If $F(p, \Omega) = F(-p, -\Omega)$, then (6.1.7) is satisfied for $n = 2$. However, whether or not (6.1.7) is satisfied for $n \geq 3$ depends on the particular form of $F(p, \Omega)$, i.e. on the original equation being studied.

6.2 Bilinear form of the Vakhnenko-Parkes equation

In order to find soliton solutions to the VPE

$$W_{XXT} + (1 + W_T)W_X = 0 \tag{6.2.1}$$

by using Hirota's method [35], we need to express it in Hirota form [74]. This is achieved by taking

$$W = 6(\ln f)_X. \tag{6.2.2}$$

Then we find that

$$W_X = \frac{3D_X^2 f \cdot f}{f^2} \quad \text{and} \quad W_{XXT} + W_X W_T = \frac{3D_T D_X^3 f \cdot f}{f^2} \tag{6.2.3}$$

and so the bilinear form of the VPE is

$$F(D_X, D_T) f \cdot f = 0, \quad F(D_X, D_T) := D_T D_X^3 + D_X^2. \tag{6.2.4}$$

The solution procedure for the VPE is to solve (6.2.4) for f by using Hirota's method and hence to find the explicit solution $W(X,T)$ for the VPE (6.2.1) by using (6.2.2). We can find $U(X,T)$ from the relation $U = W_X$ and then, as shown in Section 5.1, the implicit solution $u(x,t)$ to the VE (5.1.1) is given in parametric form by

$$u(x,t) = U(t,T), \quad x = \theta(t,T), \tag{6.2.5}$$

where

$$\theta(X,T) = T + W(X,T) + x_0. \tag{6.2.6}$$

In passing we note that the Hirota–Satsuma equation (HSE) for shallow water waves [100]

$$-u_t + u_{xxt} + uu_t + u_x \int_{-\infty}^{x} u_t \, dx' + u_x = 0 \tag{6.2.7}$$

may be written as

$$W_{xxt} + (1 + W_t)W_x - W_t = 0, \quad u = W_x = 6(\ln f)_{xx} \tag{6.2.8}$$

or in bilinear form

$$(D_t D_x^3 + D_x^2 + D_t D_x) f \cdot f = 0. \tag{6.2.9}$$

Clearly (6.2.8) and (6.2.9) are similar to, but cannot be transformed into, (5.1.10) and (6.2.4), respectively. Hence solutions to the HSE cannot be transformed into solutions of the VPE. The solution to the HSE by the Hirota method is given in [100].

6.3 Bäcklund transformation for the VPE

In Section 6.2 we wrote the VPE (5.1.10)

$$W_{XXT} + (1 + W_T) W_X = 0$$

in the Hirota bilinear form (6.2.4)

$$(D_T D_X^3 + D_X^2) f \cdot f = 0.$$

It turns out that a Bäcklund transformation follows from the bilinear form of the nonlinear evolution equation [84].

The definition of a Bäcklund transformation which was given by Rund in [113] is now the generally accepted one. Let $u(x,t)$ and $\tilde{u}(x,t)$ satisfy the partial differential equations $E(u) = 0$ and $D(\tilde{u}) = 0$, respectively. Then the set of relations $R_i((u),(\tilde{u}),(\zeta)) = 0$ $(j = 1,\ldots,n)$, where (u) and $(\tilde{u})$ denote strings, not necessarily of equal length, consisting of u, $\tilde{u}$ and their various partial derivatives, is called a Bäcklund transformation if these relations ensure that $\tilde{u}$ satisfies $D(\tilde{u}) = 0$ whenever u satisfies $E(u) = 0$ and vice versa. If u and $\tilde{u}$ satisfy the same equation, the adjective "auto" is inserted in front of Bäcklund transformation.

The main significance of Bäcklund transformations is that they have typically associated nonlinear superposition principles whereby infinite sequences of solutions to nonlinear equations may be generated by purely algebraic procedures. A Bäcklund transformation achieves the passage between different solution types, whether it is a one-soliton, two-soliton, bound state, etc. Multi-soliton solutions of many important nonlinear evolution equations can thereby be constructed. We will show that a special form of the Bäcklund transformation suggested by Hirota [114] is a key for finding an infinite number of conservation laws as well as allowing one to formulate the inverse scattering problem.

Thus, the next step in the investigation of nonlinear evolution equations should be directed to obtaining the bilinear form of the Bäcklund transformation from the bilinear form of the nonlinear equation.

6.3.1 Bäcklund transformation in bilinear form

Now we present a Bäcklund transformation for the VPE (5.1.10) written in the bilinear form (6.2.4). This type of Bäcklund transformation was first introduced by Hirota [114] and has the advantage that the transformation equations are linear with respect to each dependent variable. This Bäcklund transformation is easily transformed to the ordinary one.

We follow the method developed in [114] and described in [84]. First we define P as follows:

$$P\colon = 2\left\{[(D_T D_X^3 + D_X^2)f' \cdot f']ff - f'f'[(D_T D_X^3 + D_X^2)f \cdot f]\right\}, \tag{6.3.1}$$

where $f \neq f'$. We aim to find a pair of equations such that each equation is linear in each of the dependent variables f and f', and such that together f and f' satisfy $P = 0$. (It then follows that if f is a solution of (6.2.4) then so is f' and vice-versa.) The pair of equations is the required Bäcklund transformation.

We show that the Bäcklund transformation is given by pair of the equations

$$(D_X^3 - \lambda)f' \cdot f = 0, \tag{6.3.2}$$

$$(3D_X D_T + 1 + \mu D_X)f' \cdot f = 0, \tag{6.3.3}$$

where $\lambda = \lambda(X)$ is an arbitrary function of X and $\mu = \mu(T)$ is an arbitrary function of T.

We prove that together f and f', as determined by Eqs. (6.3.2) and (6.3.3), satisfy $P = 0$ as follows. By using the identities (VII.3), (VII.4) and Eq. (5.86) from [35] we may express P in the form

$$\begin{aligned} P =& D_T[(D_X^3 f' \cdot f) \cdot (f'f) - 3(D_X^2 f' \cdot f) \cdot (D_X f' \cdot f)] \\ &+ D_X[3(D_T D_X^2 f' \cdot f) \cdot (f'f)] - 6(D_X D_T f' \cdot f) \cdot (D_X f' \cdot f) \\ &- 3(D_X^2 f' \cdot f) \cdot (D_T f' \cdot f) + 4(D_X f' \cdot f) \cdot (f'f)]. \end{aligned} \tag{6.3.4}$$

We can rewrite P in the form

$$\begin{aligned} P =& 4D_T(\{D_X^3 - \lambda(X)\}f' \cdot f) \cdot (f'f) \\ & - 4D_X(\{3D_T D_X + 1 + \mu(T)D_X\}f' \cdot f) \cdot (D_X f' \cdot f) \end{aligned} \tag{6.3.5}$$

if we use the following identities:

$$D_X^3\left[(D_T f' \cdot f) \cdot (f f')\right] = \tag{6.3.6}$$

$$= D_T\left[(D_X^3 f' \cdot f) \cdot (f f') - 3(D_X^2 f' \cdot f) \cdot (D_X f' \cdot f)\right],$$

$$4D_T(D_X^2 f' \cdot f) \cdot (D_X f' \cdot f) = \tag{6.3.7}$$

$$= D_X[(D_T D_X^2 f' \cdot f) \cdot (f'f) + 2(D_T D_X f' \cdot f) \cdot (D_X f' \cdot f)-$$

$$-(D_X^2 f' \cdot f) \cdot (D_T f' \cdot f)] - D_X^3(D_T f' \cdot f) \cdot (f'f).$$

Identities (6.3.6) and (6.3.7) come from

$$\begin{aligned} &\exp(D_1)[\exp(D_2)f' \cdot f] \cdot [\exp(D_3)f' \cdot f] = \\ &= \exp(\tfrac{1}{2}\{D_2 - D_3\})\left[\exp\left\{\tfrac{1}{2}(D_2 + D_3) + D_1\right\} f' \cdot f\right] \\ &\qquad \& \cdot \left[\exp\left\{\tfrac{1}{2}(D_2 + D_3) - D_1\right\} f' \cdot f\right] \end{aligned} \tag{6.3.8}$$

which is Eq. (5.83) in [35], where $D_i := \varepsilon_i D_X + \delta_i D_T$. In the order $\varepsilon_1^3 \delta_3$, (6.3.8) yields (6.3.6), and in the order $\delta_1 \varepsilon_2^2 \varepsilon_3$, (6.3.8) yields (6.3.7). From (6.3.5) it follows that if (6.3.2) and (6.3.3) hold then $P = 0$ as required.

Thus we have proved that the pair of Eqs. (6.3.2) and (6.3.3) constitute a Bäcklund transformation in bilinear form for Eq. (6.2.4). Separately these equations appear as part of the Bäcklund transformation for other nonlinear evolution equations. For example, Eq. (6.3.2) is the same as one of the equations that is part of the Bäcklund transformation for a higher order KdV equation (see Eq. (5.139) in [35]), and Eq. (6.3.4) is similar to (5.132) in [35] that is part of the Bäcklund transformation for a model equation for shallow water waves.

The inclusion of μ in the operator $3D_T + \mu$ which appears in (6.3.6) corresponds to a multiplication of f and f' by terms of the form $e^{g(T)}$ and $e^{g'(T)}$, respectively; however, this has no effect o n W or W' because, from (6.2.2), $W = 6(\ln f)_X$. Hence, without loss of generality, we may take $\mu = 0$ in Eq. (6.3.3) if we wish.

6.3.2 Bäcklund transformation in ordinary form

Following the procedure given in [35, 115], we can rewrite the Bäcklund transformation in ordinary form in terms of the potential $W = \int_{-\infty}^{X} U\, dX'$ which arises as a result (5.1.8). In new variables defined by

$$\phi = \ln f'/f, \qquad \rho = \ln f'f, \tag{6.3.9}$$

Eqs. (6.3.2) and (6.3.3) have the form

$$\phi_{XXX} + 3\phi_X \rho_{XX} + \phi_X^3 - \lambda = 0, \tag{6.3.10}$$

$$3\left(\rho_{XT} + \phi_X \phi_T\right) + 1 + \mu\phi_X = 0, \tag{6.3.11}$$

respectively, where we have used results similar to (XI.1)–(XI.3) in [35]. From the definitions (6.2.2) and (6.3.9), different solutions W, W' of the VPE (5.1.10) are related to ϕ and ρ by

$$W' - W = 6\phi_X, \qquad W' + W = 6\rho_X. \tag{6.3.12}$$

Substitution of (6.3.12) into (6.3.10) and (6.3.11) with $\mu = 0$ leads to

$$(W'-W)_{XX} + \tfrac{1}{2}(W'-W)(W'+W)_X + \tfrac{1}{36}(W'-W)^3 - 6\lambda = 0, \tag{6.3.13}$$

$$\begin{aligned} &3\lambda(W' - W)_T + [(1 - W_T)((W' - W)_X + \\ &\tfrac{1}{6}(W' - W)^2) - W_{XT}(W' - W)]_X = 0, \end{aligned} \tag{6.3.14}$$

respectively. The required Bäcklund transformation in ordinary form is given by the equations (6.3.13) and (6.3.14).

Thus, by using the VPE as an example, we have traced how the bilinear and ordinary forms of the Bäcklund transformation can be found from the bilinear form of an evolution equation.

6.4 The infinite sequence of conservation laws

An important property of a soliton equation is that it has conservation laws. The existence of an infinite number of conserved quantities is associated with the intergrability of an equation [29].

A systematic way to derive higher conservation laws via the Bäcklund transformation has been developed by Satsuma; he applied it to the KdV equation [116]. Later Satsuma and Kaup [115] applied the method to a higher order KdV equation. Following [116], from the Bäcklund transformation we now construct the recurrence formula which gives the infinite sequence of conserved quantities for the VPE. An infinite sequence of conservation laws having the form

$$\frac{\partial I_n}{\partial T} + \frac{\partial F_n}{\partial X} = 0 \tag{6.4.1}$$

provides, in most cases, a corresponding sequence of integrals of motion given by the functionals $\int I_n dX$. Let us rewrite (6.3.13) (one of the Bäcklund transformation equations) in the form

$$W' - W = 6\zeta \sqrt[3]{1 - \frac{1}{6\zeta^3}\left((W' - W)_{XX} + \tfrac{1}{2}(W' - W)(W' + W)_X\right)}, \tag{6.4.2}$$

$$\zeta^3 = \lambda.$$

Assuming $1/|\zeta|$ is small, we may consider Eq. (6.4.2) to be an infinitesimal transformation from W to W'. Indeed, in the first approximation $W' \simeq W + 6\zeta$ and the next approximation with respect to $|\zeta|^{-1}$

$$W' = W + 6\zeta + \frac{1}{6\zeta} I_1.$$

Thus, we put W in the form

$$W' = W + 6\zeta + \sum_{n=1}^{\infty} \frac{1}{6^n \zeta^n} I_n(W, W_X, W_{XX}, \ldots). \tag{6.4.3}$$

Substituting Eq. (6.4.3) into Eq. (6.3.13), and equating the coefficients for the higher powers of $1/|\zeta|$, we have

$$\begin{aligned}
&O(\zeta^1): \quad I_1 = -2W_X, \\
&O(\zeta^0): \quad I_2 = 2W_{XX}, \\
&O(\zeta^{-1}): \quad I_3 = -\tfrac{4}{3}W_{XXX}, \\
&O(\zeta^{-2}): \quad I_4 = \tfrac{2}{3}W_{XXXX}, \\
&O(\zeta^{-3}): \quad I_5 = -\tfrac{2}{9}W_{XXXXX} + \tfrac{1}{9}(W_X^2)_{XX} - \tfrac{2}{9}W_{XX}^2 + \tfrac{2}{27}W_X^3.
\end{aligned} \tag{6.4.4}$$

The general recursion relations for $n \geq 5$ are as follows:

$$\begin{aligned} I_n &= -\frac{1}{3} I_{n-2,XX} - I_{n-1,X} - \frac{1}{6}\sum_{i=1}^{n-3} I_i I_{n-i-2,X} - \frac{1}{3} W_X I_{n-2} \\ &\quad -\frac{1}{6}\sum_{i=1}^{n-2} I_i I_{n-i-1} - \frac{1}{108}\sum_{i+j+l=n-2} I_i I_j I_l. \end{aligned} \tag{6.4.5}$$

The fact that these quantities are the conserved densities can be shown as follows. Let us calculate the integral $\int (W' - W - 6\zeta)_T \, dX$ with suitable boundary conditions. Taking into account Eq. (6.4.3), we have

$$\left(\int (W' - W - 6\zeta)\, dX\right)_T = \left[\frac{1}{6^n \zeta^n}\sum_{i=1}^{\infty}\left(\int I_n \, dX\right)\right]_T = 0. \tag{6.4.6}$$

Thus we deduce that the VPE has an infinite sequence of conservation laws.

6.5 The N-soliton solution for the VPE

The Hirota method for the VPE can be applied successfully if we can prove the 'N-soliton condition' (NSC) (6.1.7)–(6.1.9) for Eq. (6.2.4). Let us present this proof [75].

6.5.1 The 'N-soliton condition' for the VPE

Since for Eq. (6.2.4), we have $F(p, \Omega) = F(-p, -\Omega)$, then (6.1.7) is satisfied for $n = 2$.

With F given by (6.2.4), the dispersion relations give $\Omega_i = -1/p_i$ and (6.1.7) may be written

$$G^{(n)}(p_1, \ldots, p_n) := \tag{6.5.1}$$

$$:= \left(\prod_{i=1}^{n} p_i\right) \sum_{\sigma=\pm 1}\left\{\left(\sum_{i=1}^{n}\sigma_i p_i\right)^2 \left[1 - \left(\sum_{i=1}^{n}\frac{\sigma_i}{p_i}\right)\left(\sum_{i=1}^{n}\sigma_i p_i\right)\right] \times \right.$$

$$\left. \times \prod_{i>j}^{(n)} (\sigma_i p_i - \sigma_j p_j)^2 (p_i^2 + p_j^2 - \sigma_i \sigma_j p_i p_j)\right\}.$$

The presence of the first product term in (6.5.1) ensures that $G^{(n)}$ is a homogeneous polynomial in the p_i.

In passing we remark that previous work suggests, but does not prove, that (6.2.4) does have an NSS for all $N \geq 1$. The expression for F given by (6.2.4) is a special case of one proposed by Ito (see Eq. (B.10) in [117]). Ito claimed that this F satisfies the 3SC. Hietarinta [118] performed a search for bilinear equations of the form (6.2.4) that have an F that satisfies the 3SC. One such F was found to be the one given by (6.2.4). Hietarinta [119] later claimed that this F also passed the 4SC. The bilinear equation (6.2.4) is a special case of one given in Grammaticos et al. (see Eq. (4.4) in [120]); they showed that this equation has the Painlevé property. According to Hietarinta [119] a bilinear equation that has a 4SS and the Painlevé property is almost certainly integrable. All this evidence suggests that it is highly likely that (6.2.4) does have an NSS for all $N \geq 1$. Here we remove any doubt by using induction to prove that the condition (6.1.7) is satisfied with $G^{(n)}$ given by (6.5.1).

We need the following properties of $G^{(n)}$ (as given by (6.5.1)) for $n \geq 3$:

(i) $G^{(n)}(p_1, \ldots, p_n)|_{p_1=0} = 0$,

(ii) $$G^{(n)}(p_1, \ldots, p_n)|_{p_1=\pm p_2} = \pm 24p_1^6 \left[\prod_{i=3}^{n} (p_i^2 - p_1^2)^2 (p_i^4 + p_1^4 + p_i^2 p_1^2) \right] G^{(n-2)}(p_3, \ldots, p_n),$$

(iii) $$G^{(n)}(p_1, \ldots, p_n)|_{p_1^2+p_2^2 \pm p_1 p_2=0} = \pm(p_1 \mp p_2)(p_1^2 - p_2^2)(p_1^2 + p_2^2 \mp p_1 p_2) \times \left[\prod_{i=3}^{n} [\{(p_1 \pm p_2)^2 + p_i^2\}^2 - (p_1 \pm p_2)^2 p_i^2] \right] G^{(n-1)}(p_1 \pm p_2, p_3, \ldots, p_n).$$

(We established property (iii) by adapting the argument used to obtain equation (28) in [121] in the context of a shallow water wave equation.) Furthermore, because of the σ summation in (6.5.1), $G^{(n)}$ is an odd, symmetric function of the p_i. As already noted, the condition (6.1.7) is satisfied for $n = 1$ and $n = 2$. We now assume that the condition is satisfied for all $n \leq m - 1$, where $m \geq 3$; then the properties of $G^{(n)}$ imply that it may be factorised as follows:

$$G^{(m)}(p_1,\ldots,p_m) = \tag{6.5.2}$$

$$\left[\prod_{i=1}^{m} p_i\right]\left[\prod_{i>j}^{(m)}(p_i^2-p_j^2)^2(p_i^2+p_j^2+p_ip_j)(p_i^2+p_j^2-p_ip_j)\right]$$

$$\times\tilde{G}^{(m)}(p_1,\ldots,p_m),$$

where $\tilde{G}^{(m)}$ is a homogeneous polynomial. It follows that the degree of $G^{(m)}$ is at least $4m^2-3m$. On the other hand, from (6.5.1) the degree of $G^{(m)}$ is at most $2m^2-m+2$. As $4m^2-3m>2m^2-m+2$ for $m\geq 3$, it follows that $G^{(m)}\equiv 0$. It now follows by induction that the NSC is satisfied.

6.5.2 The N-soliton solution

With F given by (6.2.4) for the VPE

$$W_{XXT}+(1+W_T)W_X=0,$$

the dispersion relations (6.1.6) $F(2k_i,-2\omega_i)=0$ $(i=1,\ldots,N)$ give $\omega_i=1/4k_i$ and then

$$\eta_i=k_i(X-c_iT)+\alpha_i \quad \text{with} \quad c_i=1/4k_i^2. \tag{6.5.3}$$

Also, without loss of generality, we may take $k_1<\cdots<k_N$ and then

$$b_{ij}=\frac{k_j-k_i}{k_i+k_j}\sqrt{\frac{k_i^2+k_j^2-k_ik_j}{k_i^2+k_j^2+k_ik_j}}, \quad \text{where} \quad i<j, \tag{6.5.4}$$

so that $0<b_{ij}<1$.

Consequently, the relationship (6.1.4) with (6.5.3) and (6.5.4) gives f for the VPE. Finally, substitution of (6.1.4) into (6.2.2) gives the N-soliton solution $W(X,T)$ of the VPE [75].

However, following Moloney & Hodnett [122], it is more convenient to express f in the form

$$f=h_i+\hat{h}_ie^{2\eta_i} \tag{6.5.5}$$

for a given i with $1\leq i\leq N$, where

$$h_i=\sum_{\mu=0,1}\exp\left[2\left(\sum_{\substack{r=1\\(r\neq i)}}^{N}\mu_r\eta_r+\sum_{\substack{r<s\\(r\neq i,s\neq i)}}^{(N)}\mu_r\mu_s\ln b_{rs}\right)\right], \tag{6.5.6}$$

$$\hat{h}_i = \sum_{\mu=0,1} \exp\Bigg[2 \quad \Bigg(\sum_{\substack{r=1 \\ (r\neq i)}}^{N} \mu_r \eta_r + \sum_{\substack{r<s \\ (r\neq i, s\neq i)}}^{(N)} \mu_r \mu_s \ln b_{rs}$$

$$+ \sum_{r=1}^{i-1} \mu_r \ln b_{ri} + \sum_{r=i+1}^{N} \mu_r \ln b_{ir}\Bigg)\Bigg]. \tag{6.5.7}$$

Then we may write the N-soliton solution for the VPE (5.1.10) in the form

$$W(X,T) = \sum_{i=1}^{N} W_i, \tag{6.5.8}$$

$$\text{where} \quad W_i = 6k_i(1 + \tanh g_i), \quad g_i(X,T) = \eta_i + \tfrac{1}{2}\ln\left[\frac{\hat{h}_i}{h_i}\right].$$

From (6.5.8) and the relationship $U = W_X$, the corresponding expression for $U(X,T)$ is

$$U(X,T) = \sum_{i=1}^{N} U_i, \quad \text{where} \quad U_i = 6k_i \frac{\partial g_i}{\partial X} \operatorname{sech}^2 g_i. \tag{6.5.9}$$

From Section 6.2, it follows that the Moloney–Hodnett form of the N loop soliton solution to the VE (2.3.2) is given in parametric form by

$$u(x,t) = U(t,T), \quad x = \theta(t,T), \quad \theta(X,T) = T + W(X,T) + x_0 \tag{6.5.10}$$

with W and U given by (6.5.8) and (6.5.9), respectively.

The Moloney–Hodnett formulation provides a convenient way of tracking individual solitons (see Sections 6.5.4 and 6.5.5, for example).

6.5.3 The one loop soliton solution for the VE

The solution to (6.2.4) corresponding to one soliton is given by

$$f = 1 + e^{2\eta}, \quad \text{where} \quad \eta = kX - \omega T + \alpha, \tag{6.5.11}$$

and k, ω and α are constants. The dispersion relation (6.1.6) is $F(2k, -2\omega) = 0$ from which we find that $\omega = 1/4k$ and then

$$\eta = k(X - cT) + \alpha \quad \text{with} \quad c = 1/4k^2. \tag{6.5.12}$$

Substitution of (6.5.11) into (6.2.2) gives

$$W(X,T) = 6k(1+\tanh\eta) \tag{6.5.13}$$

so that

$$U(X,T) = 6k^2\operatorname{sech}^2\eta. \tag{6.5.14}$$

The one loop soliton solution to the VE is given by (6.5.10) with (6.5.13) and (6.5.14). From (6.5.10) with $v = 1/c$ we have

$$x - vt = -v(X-cT) + 6k(1+\tanh[k(X-cT)+\alpha]) + x_0. \tag{6.5.15}$$

Clearly, from (6.5.14) and (6.5.15), $U(X,T)$ and $x - vt$ are related by the parameter $\chi = X - cT$ so that $u(x,t)$ is a soliton that travels with speed v in the positive x-direction. That this soliton is a loop may be shown as follows. From (5.1.5) we have $u_x = \phi^{-1}U_T$, and on using (6.5.12) and (6.5.14) we also have $\phi = 1 - cU$ and $U_T = -cU_X$. Hence

$$u_x = -cU_X/(1-cU). \tag{6.5.16}$$

Thus, as χ goes from ∞ to $-\infty$ in (6.5.15), so that $x - vt$ goes from $-\infty$ to $+\infty$, U_X changes sign once and remains finite whereas u_x given by (6.5.16) changes sign three times and goes infinite twice. The one loop soliton solution may be written in terms of the parameter χ as

$$u = \frac{3v}{2}\operatorname{sech}^2\left(\frac{\sqrt{v}\,\chi}{2}\right), \qquad x - vt = \tilde{x_0} - v\chi + 3\sqrt{v}\tanh\left(\frac{\sqrt{v}\,\chi}{2}\right) \tag{6.5.17}$$

with $v(>0)$ and $\tilde{x_0}$ arbitrary. The solution (6.5.17) is essentially the one loop soliton solution given by (3.2.14) (see [41, 42] too).

Usually it is assumed that the value α is real so that the solution $U(X, T)$ is a real function. However, a real solution is ob-tained also with $\alpha = i\pi/2 + \tilde{\alpha}$, where $\tilde{\alpha}$ is real. In this case we have $f = 1 - e^{2\eta}$ in (6.5.11). Hence, with $\alpha = i\pi/2 + \tilde{\alpha}$ ($\tilde{\alpha}$ real) in (6.5.13) and (6.5.14), we obtain alternative real solutions for W and U, namely

$$W(X,T) = 6k(1+\coth\eta), \quad \text{where} \quad \eta = k(X-cT)+\tilde{\alpha}, \tag{6.5.18}$$

and

$$U(X,T) = -6k^2\operatorname{cosech}^2\eta. \tag{6.5.19}$$

Equation (6.5.18) is equivalent to equation (34) (with k_1 written as $2k$) in [123], and (6.5.19) is discontinuous (a singular soliton).

6.5.4 The two loop soliton solution for the VE

The solution to (6.2.4) corresponding to two solitons is given by

$$f = 1+e^{2\eta_1}+e^{2\eta_2}+b^2e^{2(\eta_1+\eta_2)}, \quad \text{where} \quad \eta_i = k_iX-\omega_iT+\alpha_i, \tag{6.5.20}$$

$$b^2 = -\frac{F[2(k_1-k_2),-2(\omega_1-\omega_2)]}{F[2(k_1+k_2),-2(\omega_1+\omega_2)]}, \tag{6.5.21}$$

and k_i, ω_i and α_i are constants. The dispersion relations (6.1.6) are $F(2k_i,-2\omega_i)=0$ $(i=1,2)$ from which we find that $\omega_i = 1/4k_i$ and then

$$\eta_i = k_i(X-c_iT)+\alpha_i \quad \text{with} \quad c_i = 1/4k_i^2. \tag{6.5.22}$$

Without loss of generality we may take $k_2 > k_1$ and then

$$b = \frac{k_2-k_1}{k_2+k_1}\sqrt{\frac{k_1^2+k_2^2-k_1k_2}{k_1^2+k_2^2+k_1k_2}}\,. \tag{6.5.23}$$

Substitution of (6.5.20) into (6.2.2) gives the two soliton solution of the VPE. Following Hodnett & Moloney [122, 124] and (6.5.8), we may write $W(X,T)$ in the form

$$W = W_1+W_2, \quad \text{where} \quad W_i = 6k_i(1+\tanh g_i) \tag{6.5.24}$$

and

$$\begin{aligned} g_1(X,T) &= \eta_1 + \frac{1}{2}\ln\left[\frac{1+b^2e^{2\eta_2}}{1+e^{2\eta_2}}\right], \\ g_2(X,T) &= \eta_2 + \frac{1}{2}\ln\left[\frac{1+b^2e^{2\eta_1}}{1+e^{2\eta_1}}\right]. \end{aligned} \tag{6.5.25}$$

As in (6.5.9), it follows that U may be written

$$U = U_1+U_2, \quad \text{where} \quad U_i = 6k_i\frac{\partial g_i}{\partial X}\operatorname{sech}^2 g_i. \tag{6.5.26}$$

The two loop soliton solution to the VE is given by (6.5.10) with (6.5.24)–(6.5.26) [74].

We now consider the two loop soliton solution in more detail. First it is instructive to consider what happens in X-T space.

As $c_1 > c_2$, we have

$$X - c_2T \to \pm\infty \quad \text{as} \quad T \to \pm\infty \quad \text{with} \quad X - c_1T \quad \text{fixed}, \tag{6.5.27}$$

and

$$X - c_1T \to \mp\infty \quad \text{as} \quad T \to \pm\infty \quad \text{with} \quad X - c_2T \quad \text{fixed}. \tag{6.5.28}$$

From (6.5.25) and (6.5.26) with (6.5.27) it follows that, with $X - c_1T$ fixed,

$$\begin{aligned} U_1 &\sim 6k_1^2 \operatorname{sech}^2 \eta_1 \quad \text{as} \quad T \to -\infty, \\ U_1 &\sim 6k_1^2 \operatorname{sech}^2(\eta_1 + \ln b) \quad \text{as} \quad T \to +\infty. \end{aligned} \tag{6.5.29}$$

Similarly, from (6.5.25) and (6.5.26) with (6.5.28), with $X - c_2T$ fixed,

$$\begin{aligned} U_2 &\sim 6k_2^2 \operatorname{sech}^2(\eta_2 + \ln b) \quad \text{as} \quad T \to -\infty, \\ U_2 &\sim 6k_2^2 \operatorname{sech}^2 \eta_2 \quad \text{as} \quad T \to +\infty. \end{aligned} \tag{6.5.30}$$

Hence it is apparent that, in the limits $T \to \pm\infty$, U_1 and U_2 may be identified as individual solitons moving with speeds c_1 and c_2 respectively in the positive X-direction. In contrast to the familiar interaction of two KdV 'sech squared' solitons [125], here it is the smaller soliton that overtakes the larger one.

The shifts Δ_i of the two solitons U_1 and U_2 in the positive X-direction due to the interaction are

$$\Delta_1 = -(\ln b)/k_1 \quad \text{and} \quad \Delta_2 = (\ln b)/k_2 \tag{6.5.31}$$

respectively. As $\ln b < 0$, the smaller soliton is shifted forwards and the larger soliton is shifted backwards. Since the 'mass' of each soliton is given by $\int_{-\infty}^{\infty} U_i dX = 12k_i$, where we have used (6.5.26), and the shifts satisfy $k_1\Delta_1 + k_2\Delta_2 = 0$, 'momentum' is conserved.

Let $r := k_1/k_2$ and recall that here we are assuming that $0 < r < 1$. (From (6.5.29) and (6.5.30), r^2 is the ratio of the amplitudes of the individual smaller and larger solitons.) Note that $U_{XX}(X_{int}, T_{int}) = 0$ for $r = R = 0.53862$, where (X_{int}, T_{int}) is the centre of the interaction. For $R < r < 1$, we have $U_{XX}(X_{int}, T_{int}) > 0$ and the two-soliton solution in X-T space always has two peaks; during interaction the two humps exchange amplitudes. For $0 < r < R$, we have

$U_{XX}(X_{int}, T_{int}) < 0$ and the two humps of the individual solitons coalesce into a single hump for part of the interaction, the smaller hump appears to pass through the larger one

Now let us consider what happens in x-t space. From (6.5.10) with $v_i = 1/c_i$ we have

$$x - v_i t = -v_i(X - c_i T) + W(X, T) + x_0. \tag{6.5.32}$$

Note that in (6.5.29) taking the limits $T \to \pm\infty$ with $X - c_1 T$ fixed is equivalent to taking the limits $X \to \pm\infty$ with $X - c_1 T$ fixed; also note that $X = t$ from (5.1.12). Accordingly from (6.5.29) and (6.5.32) with $i = 1$ we see that in the limits $t \to \pm\infty$ with $X - c_1 T$ fixed, $U_1(X, T)$ and $x - v_1 t$ are related by the parameter $X - c_1 T$. Similarly, from (6.5.30) and (6.5.32) with $i = 2$, in the limits $t \to \pm\infty$ with $X - c_2 T$ fixed, $U_2(X, T)$ and $x - v_2 t$ are related by the parameter $X - c_2 T$. It follows that in the limits $t \to \pm\infty$, u_1 and u_2 may be identified as individual loop solitons moving with speeds v_1 and v_2 respectively in the positive x-direction, where $u_i(x, t) = U_i(X, T)$. As $v_2 > v_1$, the larger loop soliton overtakes the smaller loop soliton.

The shifts, δ_i, of the two loop solitons u_1 and u_2 in the positive x-direction due to the interaction may be computed from (6.5.32) as follows. From (6.5.29), as $T \to -\infty$, $U_1 = U_{1\max} = 6k_1^2$ where $X - c_1 T = -\alpha_1/k_1$; then $W_1 = 6k_1$ and, by use of (6.5.27), $W_2 = 0$. Similarly, as $T \to \infty$, $U_1 = U_{1\max} = 6k_1^2$ where $X - c_1 T = -(\alpha_1 + \ln b)/k_1$; then $W_1 = 6k_1$ and $W_2 = 12k_2$. Use of these results in (6.5.32) with $i = 1$ gives

$$\delta_1 = 4k_1 \ln b + 12k_2. \tag{6.5.33}$$

By use of (6.5.28), (6.5.30) and (6.5.32) with $i = 2$, a similar calculation yields

$$\delta_2 = -4k_2 \ln b - 12k_1. \tag{6.5.34}$$

From (6.5.34) it is found that, for $0 < r < 1$, $\delta_2 > 0$ so that the larger loop soliton is always shifted forwards by the interaction. However, for δ_1 we find that

(a) for $r_c < r < 1$, $\delta_1 < 0$ so the smaller loop soliton is shifted backwards.

(b) for $r = r_c$, where $r_c = 0.88867$ is the root of $\ln b + 3/r = 0$, $\delta_1 = 0$ so the smaller loop soliton is not shifted by the interaction;

(c) for $0 < r < r_c$, $\delta_1 > 0$ so the smaller loop soliton is shifted forwards;

At first sight it might seem that the behaviour in (b) and (c) contradicts conservation of 'momentum'. That this is not so is justified as follows. By integrating (2.3.2) with respect to x we find that $\int_{-\infty}^{\infty} u\,dx = 0$; also, by multiplying (2.3.2) by x and integrating with respect to x we obtain $\int_{-\infty}^{\infty} xu\,dx = 0$. Thus, in x-t space, the 'mass' of each soliton is zero, and 'momentum' is conserved whatever δ_1 and δ_2 may be. In particular δ_1 and δ_2 may have the same sign as in (c), or one of them may be zero as in (b).

Cases (a), (b) and (c) are illustrated in figures 6.1, 6.2 and 6.3 respectively; in these figures u is plotted against x for various values of t. For convenience in the figures, the interactions of solitons are shown in coordinates moving with speed $(v_1 + v_2)/2$.

6.5.5 Discussion of the N loop soliton solution

We now interpret the N loop soliton solution found in Section 6.5.2 in terms of individual loop solitons [75].

First it is instructive to consider what happens in X-T space. From (6.5.8) and (6.5.9) and the fact that $c_1 > \cdots > c_N$ we deduce the following behavior: with $X - c_iT$ fixed and $T \to -\infty$,

$$U_i \sim \begin{cases} 6k_1^2 \operatorname{sech}^2 \eta_1, & \text{if } i = 1, \\ 6k_i^2 \operatorname{sech}^2 \left(\eta_i + \sum_{r=1}^{i-1} \ln b_{ri} \right), & \text{if } 2 \le i \le N; \end{cases} \tag{6.5.35}$$

with $X - c_iT$ fixed and $T \to +\infty$,

$$U_i \sim \begin{cases} 6k_i^2 \operatorname{sech}^2 \left(\eta_i + \sum_{r=i+1}^{N} \ln b_{ir} \right), & \text{if } 1 \le i \le N-1, \\ 6k_N^2 \operatorname{sech}^2 \eta_N, & \text{if } i = N. \end{cases} \tag{6.5.36}$$

Hence it is apparent that, in the limits $T \to \pm\infty$, each U_i may be identified as an individual soliton moving with speed c_i in the positive X-direction. Smaller solitons overtake larger ones.

The shifts, Δ_i, of the solitons in the positive X-direction due to the interactions between the N solitons are given by

$$\begin{aligned}\Delta_1 &= -\frac{1}{k_1}\sum_{r=2}^{N}\ln b_{1r},\\ \Delta_i &= \frac{1}{k_i}\left(\sum_{r=1}^{i-1}\ln b_{ri}-\sum_{r=i+1}^{N}\ln b_{ir}\right), \quad 2\le i\le N-1, \qquad (6.5.37)\\ \Delta_N &= \frac{1}{k_N}\sum_{r=1}^{N-1}\ln b_{rN}.\end{aligned}$$

Since the 'mass' of each soliton is given by $\int_{-\infty}^{\infty} U_i\,dX = 12k_i$, where we have used (6.5.9), and the shifts satisfy

$$\sum_{i=1}^{N} k_i\Delta_i = 0, \qquad (6.5.38)$$

'momentum' is conserved.

Now let us consider what happens in x-t space. From (6.5.10) with $v_i = 1/c_i$ we have

$$x - v_i t = -v_i(X - c_iT) + W(X,T) + x_0. \qquad (6.5.39)$$

Note that in (6.5.35) and (6.5.36) taking the limits $T \to \pm\infty$ with $X - c_iT$ fixed is equivalent to taking the limits $X \to \pm\infty$ with $X - c_iT$ fixed; also note that $X = t$ from (5.1.12). Accordingly from (6.5.35), (6.5.36) and (6.5.39), with a given i, we see that in the limits $t \to \pm\infty$ with $X - c_iT$ fixed, $U_i(X,T)$ and $x - v_it$ are related by the parameter $X - c_iT$. It follows that in the limits $t \to \pm\infty$, u_i may be identified as an individual loop soliton moving with speed v_i in the positive x-direction, where $u_i(x,t) = U_i(X,T)$. As $v_1 < \cdots < v_N$, larger loop solitons overtake smaller ones.

In order to calculate the shifts, δ_i, of the loop solitons u_i in the positive x-direction due to the interactions between the N loop solitons, we need the following results: from (6.5.35), as $T \to -\infty$, $U_i \to U_{i\max} = 6k_i^2$ where

$$X - c_iT = \begin{cases} -\dfrac{\alpha_1}{k_1}, & \text{for } i = 1 \text{ and then } W \to 6k_1,\\ -\dfrac{\alpha_i}{k_i} - \dfrac{1}{k_i}\displaystyle\sum_{r=1}^{i-1}\ln b_{ri}, & \text{for } 2 \le i \le N \text{ and then } W \to 6k_i + \displaystyle\sum_{r=1}^{i-1}12k_r;\end{cases}$$

from (6.5.36), as $T \to \infty$, $U_i \to U_{i\max} = 6k_i^2$ where

$$X - c_i T = \begin{cases} -\dfrac{\alpha_i}{k_i} - \dfrac{1}{k_i} \displaystyle\sum_{r=i+1}^{N} \ln b_{ir}, \\ \qquad\qquad \text{for } 1 \le i \le N-1 \text{ and then } W \to 6k_i + \displaystyle\sum_{r=i+1}^{N} 12k_r. \\ -\dfrac{\alpha_N}{k_N}, \text{ for } i = N \text{ and then } W \to 6k_N. \end{cases}$$

Use of these results in (6.5.39) gives

$$\delta_1 = \sum_{r=2}^{N} (4k_1 \ln b_{1r} + 12k_r),$$

$$\delta_i = \sum_{r=i+1}^{N} (4k_i \ln b_{ir} + 12k_r) - \sum_{r=1}^{i-1} (4k_i \ln b_{ri} + 12k_r), \quad 2 \le i \le N-1, \tag{6.5.40}$$

$$\delta_N = -\sum_{r=1}^{N-1} (4k_N \ln b_{rN} + 12k_r).$$

Example: $N = 3$

As an example, here we consider the interaction of three solitons [75]. The interaction process is more complicated than that for the two loop soliton solution [74] given in Section 6.5.4.

For $N = 3$, (6.5.36) gives

$$f = 1 + e^{2\eta_1} + e^{2\eta_2} + e^{2\eta_2} + b_{12}^2 e^{2(\eta_1+\eta_2)} + b_{13}^2 e^{2(\eta_1+\eta_3)} + b_{23}^2 e^{2(\eta_2+\eta_3)}$$
$$+ b_{12}^2 b_{13}^2 b_{23}^2 e^{2(\eta_1+\eta_2+\eta_3)} \tag{6.5.41}$$

so that, for example, (6.5.8) gives

$$W_1 = 6k_1(1 + \tanh g_1), \tag{6.5.42}$$

where

$$g_1(X,T) = \eta_1 = \frac{1}{2} \ln \left[\frac{1 + b_{12}^2 e^{2\eta_2} + b_{13}^2 e^{2\eta_3} + b_{12}^2 b_{13}^2 b_{23}^2 e^{2(\eta_2+\eta_3)}}{1 + e^{2\eta_2} + e^{2\eta_2} + b_{23}^2 e^{2(\eta_2+\eta_3)}} \right]. \tag{6.5.43}$$

Also (6.5.40) give

$$\delta_1 = 4k_1(\ln b_{12} + \ln b_{13}) + 12(k_2 + k_3), \tag{6.5.44}$$

$$\delta_2 = 4k_2(\ln b_{23} - \ln b_{12}) + 12(k_3 - k_1), \tag{6.5.45}$$

$$\delta_3 = -4k_3(\ln b_{13} + \ln b_{23}) - 12(k_1 + k_2). \tag{6.5.46}$$

It is found that $\delta_3 > 0$ so that the largest loop soliton is always shifted forwards by the interaction with the other two loop solitons. However, δ_1 and δ_2 may be positive or negative depending on the values of the ratios $r_{12}{:} = k_1/k_2$ and $r_{23}{:} = k_2/k_3$. (Note that $0 < r_{12} < 1$ and $0 < r_{23} < 1$.) This is illustrated in figures 6.4 and 6.5, respectively. It is clear that it is possible to choose k_1, k_2 and k_3 such that δ_1, δ_2 and δ_3 are all positive. At first sight it might seem that this contradicts conservation of momentum. However, as pointed out in [74], the mass of each loop soliton is zero, and momentum is conserved whatever the values of δ_1, δ_2 and δ_3.

It is of interest to investigate the nature of the interactions between the three loop solitons. Here we consider the case where all three solitons arrive at $x = 0$ at time $t = 0$. For the interactions to be centred at $X = 0$ and $T = 0$ in X–T space, we require

$$\alpha_1 = -\tfrac{1}{2}(\ln b_{12} + \ln b_{13}),$$

$$\alpha_2 = -\tfrac{1}{2}(\ln b_{12} + \ln b_{23}),$$

$$\alpha_3 = -\tfrac{1}{2}(\ln b_{13} + \ln b_{23}),$$

and

$$x_0 = -6(k_1 + k_2 + k_3).$$

Before discussing the interaction process it is helpful to recall some results for the two loop soliton solution given in [74]. It was shown that there are three characteristic types of behaviour during the interaction process and that these depend on the value of the ratio r_{12} as follows:

(1) for $0.759\,68 < r_{12} < 1$, the two loops exchange their amplitudes during the interaction but never overlap;

(2) for $0.556\,76 < r_{12} < 0.759\,68$, the two loops exchange their amplitudes during the interaction and, for part of the interaction, the loops overlap;

(3) for $0 < r_{12} < 0.556\,76$, the larger loop catches up the smaller loop which then travels clockwise around the larger loop before being ejected behind the larger loop.

For the three loop soliton solution, the behaviour during the interaction process clearly depends on both ratios r_{12} and r_{23}. Similar types of behaviour to the above can be observed for the interaction between loops 1 and 2, and between loops 2 and 3, where we have denoted the loop corresponding to k_i as loop i. Below we consider three illustrative examples. For each example we present a figure in which u is plotted against $x - v_2 t$ at several equally spaced values of t, and give the values of δ_i ($i = 1$; 2; 3) as calculated from (6.5.44)–(6.5.46).

First, consider the case where $k_1 = 0.36$, $k_2 = 0.9$ and $k_3 = 1$ so that $r_{12} = 0.4$ and $r_{23} = 0.9$. From the results in [74] we might expect loop 1 to travel clockwise around loop 2 before being ejected behind it, and loops 2 and 3 to exchange their amplitudes but never overlap. As can be seen from figure 6.6 this is indeed what happens. In this case $\delta_1 = 19.50$, $\delta_2 = -0.54$ and $\delta_3 = 3.18$.

Second, consider the case where $k_1 = 0.35$, $k_2 = 0.5$ and $k_3 = 1$ so that $r_{12} = 0.7$ and $r_{23} = 0.5$. From the results in [74] we might expect loops 1 and 2 to overlap for part of the interaction and to exchange their amplitudes during the interaction, and for loop 2 to travel clockwise around loop 3. As can be seen from figure 6.7 not only does this happen but after loops 1 and 2 overlap, they both travel around loop 3 before both are ejected behind loop 3. In this case $\delta_1 = 13.38$, $\delta_2 = 9.24$ and $\delta_3 = 0.10$.

Third, consider the case where $k_1 = 0.405$, $k_2 = 0.45$ and $k_3 = 1$ so that $r_{12} = 0.9$ and $r_{23} = 0.45$. From the results in [74] we might expect loops 1 and 2 to exchange their amplitudes but never overlap, and loop 2 to travel clockwise around loop 3. As can be seen from figure 6.8 both loops 1 and 2 travel clockwise around loop 3 and exchange amplitudes, but they overlap for a while near and at $t = 0$. In this case $\delta_1 = 9.77$, $\delta_2 = 10.97$ and $\delta_3 = 0.08$.

Clearly many other types of interaction are possible and, as demonstrated by our third example, it is not always possible to predict what will happen on the basis of the results in [74] (see also Section 6.5.4) alone. The interaction process for the three loop soliton solution is more complicated than that for the two loop soliton solution; we have been unable to classify the interactions into distinct characteristic cases for ranges of values of the ratios r_{12} and r_{23} in a way similar to that for the two loop solution. Nevertheless, the results of Section 6.5.4 can give us a rough indication as to what might happen during the interaction process for the three loop soliton solution.

6.6 The N-soliton solution for the GVPE

6.6.1 The GVPE

In Section 6.2 we observed that the VPE

$$W_{XXT}+(1+W_T)W_X=0 \tag{6.6.1}$$

may be written in the bilinear form

$$F(D_X,D_T)f\cdot f=0, \quad \text{where} \quad F(D_X,D_T){:}=D_TD_X^3+D_X^2. \tag{6.6.2}$$

We now consider a slightly more general form of (6.6.2), namely

$$F(D_X,D_T)f\cdot f=0, \text{ where } F(D_X,D_T){:}=D_TD_X^3+D_X^2+\beta D_XD_T \tag{6.6.3}$$

and β is a free parameter. Now, on using the identities (6.2.3) together with the identity

$$W_T=\frac{3D_XD_Tf\cdot f}{f^2}, \tag{6.6.4}$$

we may write (6.6.3) as

$$W_{XXT}+(1+W_T)W_X+\beta W_T=0. \tag{6.6.5}$$

We refer to (6.6.5) as the generalised VPE (GVPE).

We note that, by using the relationships (5.1.5), (5.1.8) and (5.1.9), we may transform the GVPE into the GVE given by (4.5.5), namely

$$\frac{\partial}{\partial x}\left(\mathcal{D}^2u+\frac{1}{2}u^2+\beta u\right)+\mathcal{D}u=0. \tag{6.6.6}$$

In Section 4.5 this equation was derived in a different way via a discussion of a Hirota–Satsuma-type 'shallow water wave' equation.

6.6.2 The N-soliton solution for the GVPE

In Section 6.5.1 we discussed the 'N-soliton condition' (NSC) for the F in (6.6.2) associated with the VPE. A similar discussion for the F in (6.6.3) associated with the GVPE is given in the appendix of [101] where it is shown that the NSC is satisfied. The N-soliton solution to the GVPE is found by an appropriate generalisation of the procedure for the VPE given in Section 6.5.2.

With F given by (6.6.3) the dispersion relations (6.1.6) $F(2k_i, -2\omega_i) = 0$ $(i = 1, \ldots, N)$ give $\omega_i = k_i/(4k_i^2 + \beta)$ and then

$$\eta_i = k_i(X - c_i T) + \alpha_i \quad \text{with} \quad c_i = 1/(4k_i^2 + \beta). \tag{6.6.7}$$

Also, without loss of generality, we may take $k_1 < \cdots < k_N$ and then

$$b_{ij} = \frac{k_j - k_i}{k_i + k_j}\sqrt{\frac{4k_i^2 + 4k_j^2 - 4k_i k_j + 3\beta}{4k_i^2 + 4k_j^2 + 4k_i k_j + 3\beta}}, \quad \text{where} \quad i < j, \tag{6.6.8}$$

so that $0 < b_{ij} < 1$.

Consequently, the relationship (6.1.4) with (6.6.7) and (6.6.8) gives f for the GVPE. Finally, substitution of (6.1.4) into (6.2.2) gives the N-soliton solution $W(X, T)$ of the GVPE [101]. As described in Section 6.5.2, the solution may also be expressed in the Moloney–Hodnett form (6.5.8). The corresponding N-soliton solution $u(x, t)$ of the GVE (6.6.6) may then be expressed in parametric form given by (6.5.9) and (6.5.10).

A discussion of the properties of the N-soliton solution of the GVPE and the corresponding N-soliton solution of the GVE is given in Section 6 in [101]. The one-soliton and two-soliton solutions are discussed in more detail in Sections 4 and 7, respectively.

6.7 The N-soliton solution for the mGVPE

6.7.1 The mGVPE

In Section 4.5 we observed that the modified generalised VE (mGVE), namely

$$\frac{\partial}{\partial x}\left(\mathcal{D}^2 u + qu^2 + \beta u\right) + q\mathcal{D}u = 0, \tag{6.7.1}$$

where q and β are arbitrary non-zero constants, may be derived from a Hirota–Satsuma-type 'shallow water wave' equation of the form

$$U_{XXT}+2qUU_T-qU_X\int_X^\infty U_T(X',T)\,dX'+\beta U_T+qU_X=0, \quad (6.7.2)$$

In terms of W, where $W_X=U$, (6.7.2)becomes

$$W_{XXXT}+2qW_XW_{XT}+qW_{XX}W_T+\beta W_{XT}+qW_{XX}=0. \quad (6.7.3)$$

Equation (6.7.3) is the mGVPE. In order to put it into bilinear form, following Hirota and Satsuma [100], we introduce an auxiliary variable τ. By taking

$$W=\frac{4}{q}(\ln f)_X\,, \quad (6.7.4)$$

we find that (6.7.3) may be expressed as the two coupled bilinear equations

$$F(D_X,D_T,D_\tau)(f\cdot f)=0 \quad (6.7.5)$$

and

$$G(D_X,D_\tau)(f\cdot f)=0, \quad (6.7.6)$$

where

$$F(D_X,D_T,D_\tau):=\beta D_XD_T+D_X^3D_T+qD_X^2-\frac{1}{3}\left(D_\tau D_T+D_X^3D_T\right) \quad (6.7.7)$$

and

$$G(D_X,D_\tau):=D_X\left(D_\tau+D_X^3\right). \quad (6.7.8)$$

6.7.2 The N-soliton solution for the mGVPE

In the appendix of [102] the N-soliton conditions for F and G in (6.7.5) and (6.7.6), respectively, are shown to be satisfied. The solution procedure for the mGVPE is to solve (6.7.5) subject to (6.7.6) for f by using Hirota's method and then to find $W(X,T)$ by using (6.7.4).

The solution of (6.7.5) and (6.7.6) corresponding to N solitons is given by

$$f = \sum_{\mu=0,1} \exp\left[2\left(\sum_{i=1}^{N} \mu_i \eta_i + \sum_{i<j}^{(N)} \mu_i \mu_j \ln b_{ij}\right)\right], \tag{6.7.9}$$

where $\eta_i = k_i X - \omega_i T - d_i \tau + \alpha_i,$

$$b_{ij}^2 = -\frac{F[2(k_i - k_j), -2(\omega_i - \omega_j), -2(d_i - d_j)]}{F[2(k_i + k_j), -2(\omega_i + \omega_j), -2(d_i + d_j)]} \tag{6.7.10}$$

$$= -\frac{G[2(k_i - k_j), -2(d_i - d_j)]}{G[2(k_i + k_j), -2(d_i + d_j)]}, \tag{6.7.11}$$

and k_i, ω_i, d_i and α_i are constants. From the dispersion relations $G(2k_i, -2d_i) = 0$ $(i = 1, \ldots, N)$ we obtain $d_i = 4k_i^3$ and then, from the dispersion relations $F(2k_i, -2\omega_i, -2d_i) = 0$ $(i = 1, \ldots, N)$, we obtain $\omega_i = qk_i/(4k_i^2 + \beta)$ so that

$$\eta_i = k_i(X - c_i T) - 4k_i^3 \tau + \alpha_i \quad \text{with} \quad c_i = q/(4k_i^2 + \beta). \tag{6.7.12}$$

Also, without loss of generality, we may take $k_1 < \cdots < k_N$ and then, using either (6.7.10) or (6.7.11),

$$b_{ij} = \frac{k_j - k_i}{k_j + k_i}, \quad \text{where} \quad i < j, \tag{6.7.13}$$

so that $0 < b_{ij} < 1$.

In principle, substitution of (6.7.9) with (6.7.12) and (6.7.13) into (6.7.4) gives the N-soliton solution $W(X,T)$ for the mGVPE. The solution may also be expressed in Moloney–Hodnett form. It is

$$W(X,T) = \sum_{i=1}^{N} W_i, \tag{6.7.14}$$

where $W_i = \dfrac{4}{q} k_i (1 + \tanh g_i), \quad g_i(X,T) = \eta_i + \tfrac{1}{2} \ln\left[\dfrac{\hat{h}_i}{h_i}\right],$

and $\hat{h}_i$ and h_i are given by (6.5.6) and (6.5.7), respectively. The corresponding N-soliton solution $u(x,t)$ for the mGVE (6.7.1) may then be expressed in parametric form given by

$$U(X,T) = \sum_{i=1}^{N} U_i, \quad \text{where} \quad U_i = \frac{4}{q} k_i \frac{\partial g_i}{\partial X} \operatorname{sech}^2 g_i, \tag{6.7.15}$$

together with (6.5.10).

A discussion of the properties of the N-soliton solution of the mGVPE and the corresponding N-soliton solution of the GVE is given in Section 6 in [102]. The one-soliton and two-soliton solutions are discussed in more detail in Sections 4 and 7, respectively.

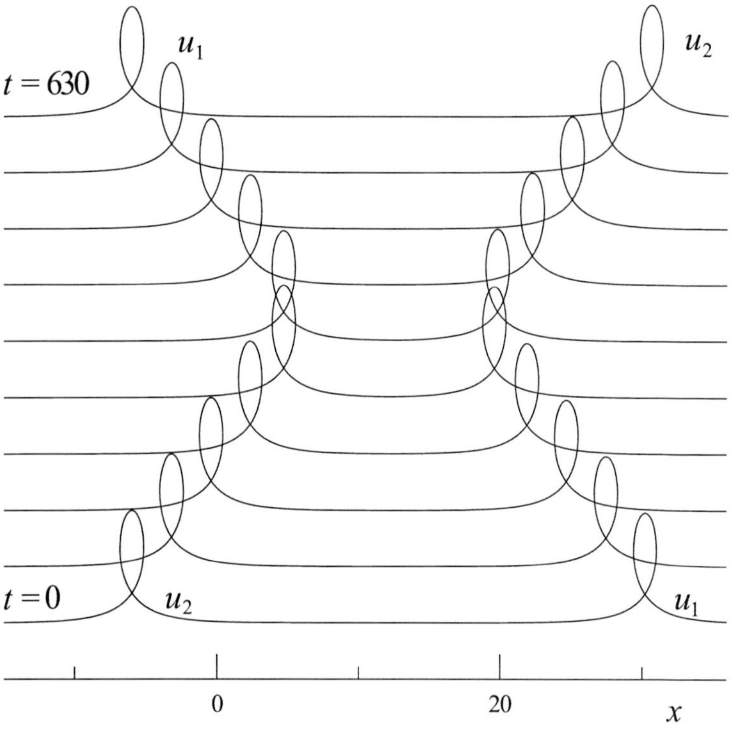

Figure 6.1: The interaction process for two loop solitons with $k_1 = 0.99$ and $k_2 = 1$, so that $r = 0.99$ and $\delta_1 < 0$.

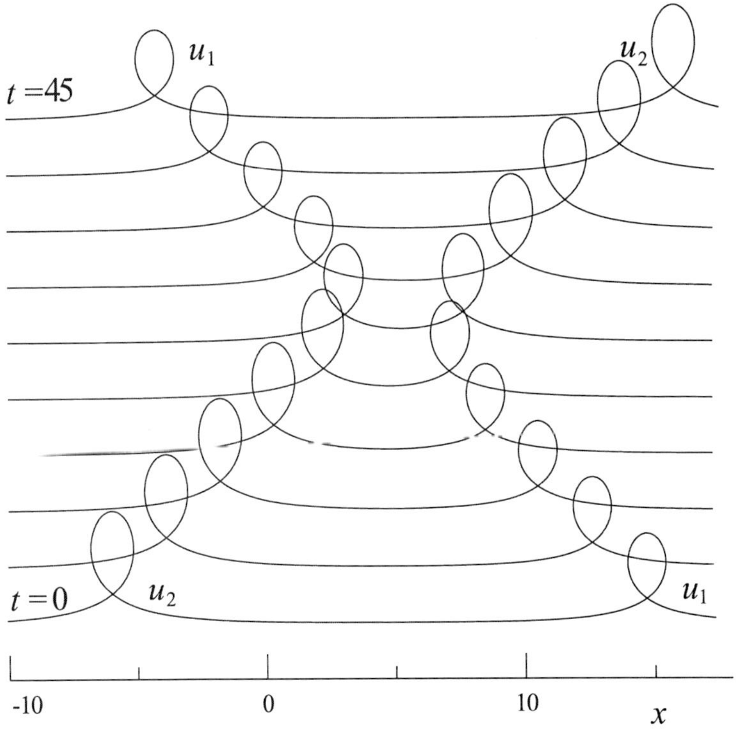

Figure 6.2: The interaction process for two loop solitons with $k_1 = 0.88867$ and $k_2 = 1$, so that $r = 0.88867$ and $\delta_1 = 0$.

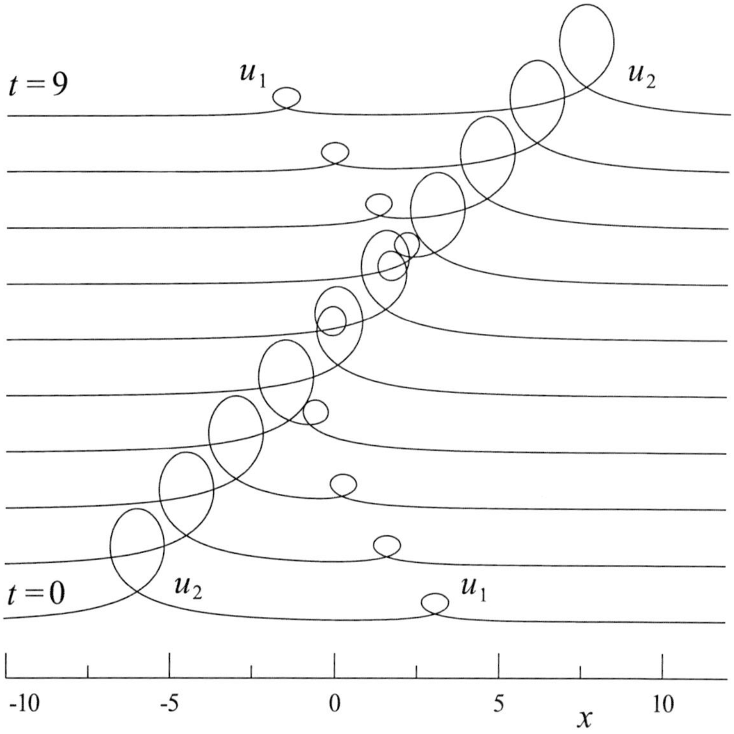

Figure 6.3: The interaction process for two loop solitons with $k_1 = 0.5$ and $k_2 = 1$, so that $r = 0.5$ and $\delta_1 > 0$.

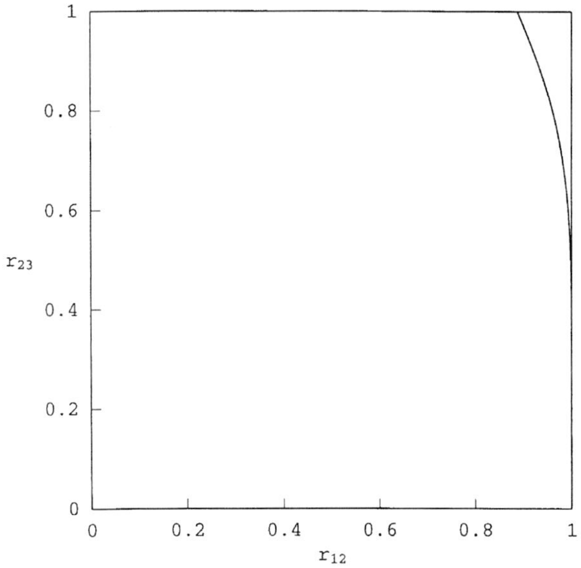

Figure 6.4: δ_1 as given by (6.5.44) is positive in the larger region and negative in the smaller region.

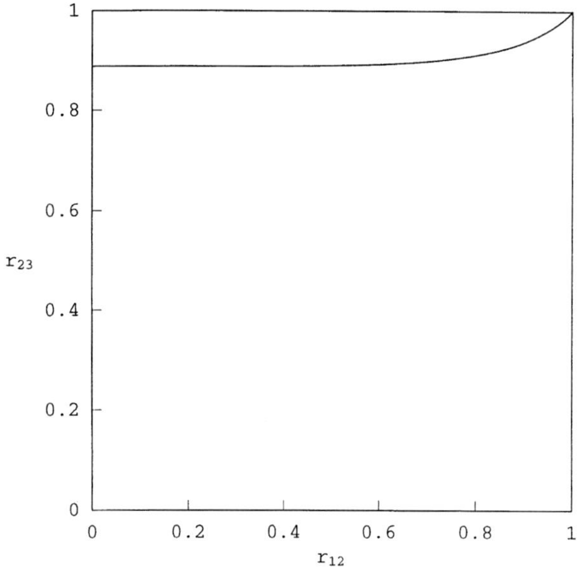

Figure 6.5: δ_2 as given by (6.5.45) is positive in the larger region and negative in the smaller region.

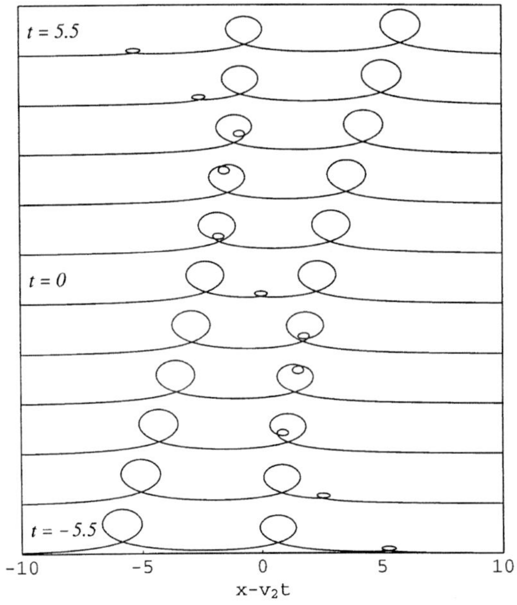

Figure 6.6: The interaction process for three loop solitons with $r_{12} = 0.4$ and $r_{23} = 0.9$.

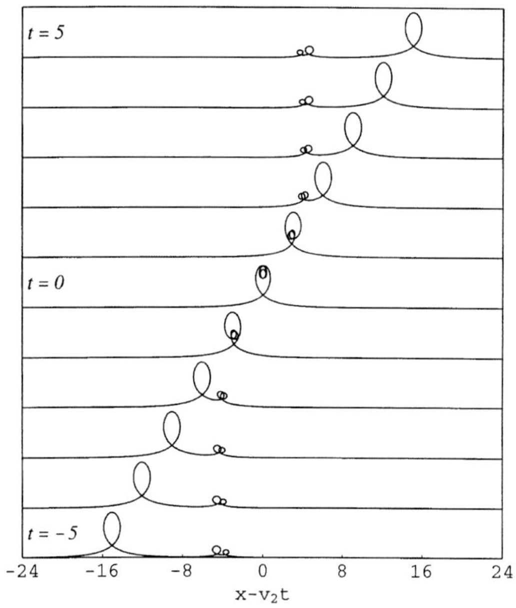

Figure 6.7: The interaction process for three loop solitons with $r_{12} = 0.7$ and $r_{23} = 0.5$.

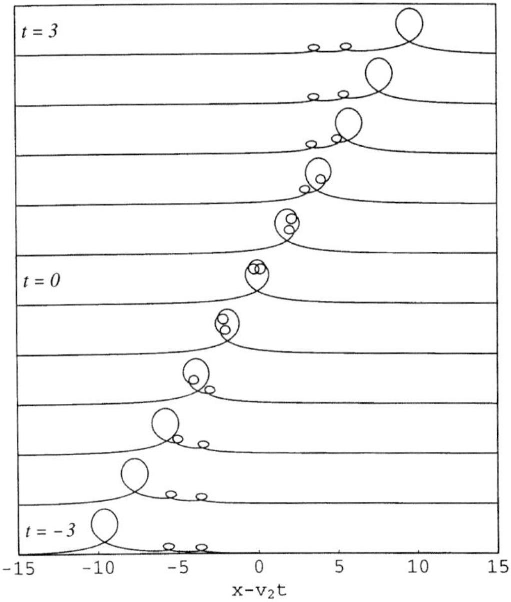

Figure 6.8: The interaction process for three loop solitons with $r_{12} = 0.9$ and $r_{23} = 0.45$.

Chapter 7
The inverse scattering method

The inverse scattering transform (IST) method is arguably the most important discovery in the theory of solitons. The method enables one to solve the initial value problem for a nonlinear evolution equation. Moreover, it provides a proof of the complete integrability of the equation.

The idea of the inverse scattering method was first introduced for the KdV equation [94] and subsequently developed for the nonlinear Schrödinger equation [28], the mKdV equation [126, 127], the sine-Gordon equation [25, 128] and the equation of motion for a one-dimensional exponential lattice (Toda lattice) [129]. It is to be remarked that the inverse method is a unique theory whereby the initial value problem for the nonlinear differential equations can be solved exactly. For the KdV equation this method was expressed in general form by Lax [130].

The essence of the application of the IST is as follows. The equation of interest for study (in our case the VPE (5.1.10)) is written as the compatibility condition for two linear equations. These equations, (7.1.2) and (7.1.3), will be derived below. Then $W(X,0)$ is mapped into the scattering data $S(0)$ for (7.1.2). It is important that, since the variable $W(X,T)$ contained in the spectral equation (7.1.2) evolves according to Eq. (5.1.10), the spectrum λ always retains constant values. The time evolution of $S(T)$ is simple and linear. From a knowledge of $S(T)$, we may reconstruct $W(X,T)$.

7.1 Formulation of the inverse scattering eigenvalue problem

Since we have obtained the N-loop soliton solution to the VPE (5.1.10) by use of the Hirota method (see Section 6.5), we can state that the VPE (5.1.10) is integrable. The use of the IST is the most appropriate way of tackling the initial value problem. In order to apply the IST method, one first has to formulate the associated eigenvalue problem. This can be achieved by finding a Bäcklund transformation associated with the VPE. We have already shown in Section 6.3 that the Bäcklund transformation is one of the analytical tools for dealing with soliton problems. The main aim of Section 7.1 is to give the details of the IST method for solving the VPE, so first we will formulate the scattering problem.

Now we will show that the IST problem for the VPE in the form (5.1.10) has a third-order eigenvalue problem that is similar to the one associated with a higher order KdV equation [115, 131], a Boussinesq equation [131, 132, 133, 134, 135], and a model equation for shallow water waves [96, 35].

Introducing the function

$$\psi = f'/f, \tag{7.1.1}$$

and taking into account (6.2.1), we find that (6.3.2) and (6.3.3) reduce to

$$\psi_{XXX} + W_X\psi_X - \lambda\psi = 0, \tag{7.1.2}$$
$$3\psi_{XT} + (1 + W_T)\psi + \mu\psi_X = 0, \tag{7.1.3}$$

respectively, where we have used results similar to (X.1) – (X.3) in [35]. Recall that $\lambda = \lambda(X)$ is an arbitrary function of X and $\mu = \mu(T)$ is an arbitrary function of T.

From (7.1.2) and (7.1.3) it can be shown that

$$3\lambda\psi_T + (1+W_T)\psi_{XX} - W_{XT}\psi_X + [W_{XXT} + (1+W_T)W_X + \mu\lambda]\psi = 0 \tag{7.1.4}$$

and

$$[W_{XXT} + (1 + W_T)W_X]_X\psi + (3\psi_T + \mu\psi)\lambda_X = 0. \tag{7.1.5}$$

In view of (5.1.10), (7.1.4) becomes

$$3\lambda\psi_T + (1 + W_T)\psi_{XX} - W_{XT}\psi_X + \lambda\mu\psi = 0, \tag{7.1.6}$$

and (7.1.5) implies that $\lambda_X = 0$ so the spectrum λ of (7.1.2) remains constant. Constant λ is what is required in the IST problem. Note that, with constant λ, equation (7.1.5) yields the equation

$$W_{XXT} + (1 + W_T)W_X = h(T),$$

where $h(T)$ is an arbitrary function of T. Now, according to (7.1.17) and (7.1.30), the inverse scattering method restricts the solutions to those that vanish as $|X| \to \infty$, so $h(T)$ is to be identically zero. Thus the pair of equations (7.1.2), (7.1.3) or (7.1.2) and (7.1.5) can be considered as the Lax pair for the VPE (5.1.10).

Since (7.1.2) and (7.1.3) are alternative forms of Eqs. (6.3.2) and (6.3.3), respectively, it follows that the pair of equations (7.1.2) and (7.1.3) is associated with the VPE (5.1.10) considered here. Thus the IST problem is directly related to a spectral equation of third order, namely (7.1.2). As for the VPE, the third-order spectral equation is associated with a Boussinesq equation [132, 131, 133, 140, 134, 135], a higher order KdV equation [131, 115], a model equation for shallow water waves [96, 35]. The inverse problem for certain third-order spectral equations has been considered by Kaup [131] and Caudrey [132, 133]. As expected, (7.1.2) and (7.1.3) are similar to, but cannot be transformed into, the corresponding equations for the Hirota–Satsuma equation (HSE) (see Eqs. (A8a) and (A8b) in [136]). Clarkson and Mansfield [137] note that the scattering problem for the HSE is similar to that for the Boussinesq equation which has been studied comprehensively by Deift et al. [135].

After the Lax pair for the VPE was derived in [84], in [73] the Lax pair was written in its original variables as a zero curvature condition. Moreover, in [73] Hone and Wang have shown that there is a subtle connection between the Sawada-Kotera hierarchy and the VE, between the Degasperis-Procesi equation (DPE) and the VE (see also [69, 70]), and between the Lax pairs of the DPE and the VE. For the Cauchy problem at long-time, the IST approach presents throughout a Riemann-Hilbert problem [138, 139] in original (physical) independent variables for the VE in [139].

7.1.1 Example of the use of the IST method to find the one-soliton solution

Here we re-derive the one-soliton solution (6.5.13) of the VPE as obtained in Section 6.5.3 by the Hirota method, but instead by application of the IST method. Let the initial perturbation be

$$W(X,0) = 6k(1+\tanh(\eta)), \quad \eta = kX + \alpha. \tag{7.1.7}$$

For convenience we introduce new notation ξ_1, β_1 instead of parameters k and α by

$$k = \tfrac{\sqrt{3}}{2}\xi_1, \quad \alpha = \tfrac{1}{2}\ln(\beta_1/2\sqrt{3}\xi_1) \tag{7.1.8}$$

then

$$W(X,0) = 6\frac{\partial}{\partial X}\ln\left[1 + \frac{\beta_1}{2\sqrt{3}\xi_1}\exp(\sqrt{3}\xi_1 X)\right] \tag{7.1.9}$$

is the initial condition for the VPE.

The first step in the IST method is to solve the spectral equation (7.1.2) with spectral parameter λ for the given initial condition $W(X,0)$. In our example it is (7.1.9). The solution is studied over the complex ζ-plane, where $\zeta^3 = \lambda$. One can verify by direct substitution of (7.1.10) in (7.1.2) that the solution $\psi(X,0;\zeta)$ of the linear ODE (7.1.2), normalized so that $\psi(X,0;\zeta)\exp(-\zeta X) \to 1$ at $X \to -\infty$, is given by

$$\psi(X,0;\zeta)\exp(-\zeta X) = \tag{7.1.10}$$

$$= 1 - \frac{\beta_1 \exp\left(\sqrt{3}\xi_1 X\right)}{1 + \beta_1 \dfrac{\exp(\sqrt{3}\xi_1 X)}{2\sqrt{3}\xi_1}}\left[\frac{\omega_2}{\mathrm{i}\omega_2\xi_1 - \zeta} + \frac{\omega_3}{-\mathrm{i}\omega_3\xi_1 - \zeta}\right],$$

where $\omega_j = e^{\mathrm{i}2\pi(j-1)/3}$ are the cube of roots of 1 ($j = 1, 2, 3$). The constants β_1 and ξ_1, as we will show, are associated with the local spectral data.

The second step in the IST method is to obtain the evolution of β_1 and ξ_1 with respect to T. The time-dependence of the solution $\psi(X,T)$ is described by equation (7.1.3). Analyzing equation (7.1.3), we may assume that

$$\begin{aligned} \xi_1(T) &= \xi_1(0) = \xi_1 = \text{const.}, \\ \beta_1(T) &= \beta_1(0)\exp\left(-\frac{1}{\sqrt{3}\xi_1}T\right) = \beta_1 \exp\left(-\frac{1}{\sqrt{3}\xi_1}T\right). \end{aligned} \tag{7.1.11}$$

Below, the assumption of these relationships will be justified. Indeed, we know that the spectrum λ in (7.1.2) remains constant if $W(X,T)$ evolves according to Eq. (5.1.10). Therefore, as will be proved, the spectrum data evolve as in (7.1.28). In the notations (7.2.5), (7.2.6), from (7.1.28) we obtain the relations (7.1.11).

The final step in the IST method is to select the solution $W(X,T)$ from (7.1.10) with $\xi_1(T)$, $\beta_1(T)$ as in (7.1.11). According to Eq. (2.7) in [131] we expand $\psi(X,T;\zeta)$ as an asymptotic series in ζ^{-1} to obtain

$$\psi(X,0;\zeta)\exp(-\zeta X) = 1-\frac{1}{3\zeta}\left[W(X)-W(-\infty)\right]+O(\zeta^{-2}), \quad (7.1.12)$$

i.e.

$$W(X)-W(-\infty) = \lim_{\zeta\to\infty}\left[3\zeta(1-\psi\exp(-\zeta X))\right].$$

Taking into account the functional dependence (7.1.11), we find the required one-soliton solution of the VPE in form

$$W(X,T) = 6\frac{\partial}{\partial X}\ln\left[1+\frac{\beta_1}{2\sqrt{3}\xi_1}\exp\left(\sqrt{3}\xi_1 X-\frac{1}{\sqrt{3}\xi_1}T\right)\right]+\text{const}. \quad (7.1.13)$$

This expression (7.1.13) may be written in the form (6.5.13). Thus, for the example of the one-soliton solution, we have demonstrated the IST method.

7.1.2 The direct spectral problem

Let us consider the principal aspects of the inverse scattering transform problem for a third-order equation. The inverse problem for certain third-order spectral equations has been considered by Kaup [131] and Caudrey [132, 133]. The time evolution of ψ is determined from (7.1.3) or (7.1.6).

Following the method described by Caudrey [132], the spectral equation (7.1.2) can be rewritten

$$\frac{\partial}{\partial X}\boldsymbol{\psi} = \left[\mathbf{A}(\zeta)+\mathbf{B}(X,\zeta)\right]\cdot\boldsymbol{\psi} \quad (7.1.14)$$

with

$$\boldsymbol{\psi} = \begin{pmatrix}\psi\\ \psi_X\\ \psi_{XX}\end{pmatrix},\ \mathbf{A} = \begin{pmatrix}0&1&0\\0&0&1\\ \lambda&0&0\end{pmatrix},\ \mathbf{B} = \begin{pmatrix}0&0&0\\0&0&0\\0&-W_X&0\end{pmatrix}. \quad (7.1.15)$$

The matrix $\mathbf{A}$ has eigenvalues $\lambda_j(\zeta)$ and left- and right-eigenvectors $\tilde{\boldsymbol{v}}_j(\zeta)$ and $\boldsymbol{v}_j(\zeta)$, respectively. These quantities are defined through a spectral parameter λ as

$$\lambda_j(\zeta) = \omega_j \zeta, \quad \lambda_j^3(\zeta) = \lambda,$$
$$\boldsymbol{v}_j(\zeta) = \begin{pmatrix} 1 \\ \lambda_j(\zeta) \\ \lambda_j^2(\zeta) \end{pmatrix}, \quad \tilde{\boldsymbol{v}}_j(\zeta) = \begin{pmatrix} \lambda_j^2(\zeta) & \lambda_j(\zeta) & 1 \end{pmatrix}, \tag{7.1.16}$$

where, as previously, $\omega_j = e^{2\pi \mathrm{i}(j-1)/3}$ are the cube roots of 1 ($j =$ 1, 2, 3). Obviously the $\lambda_j(\zeta)$ are distinct and they and $\tilde{\boldsymbol{v}}_j(\zeta)$ and $\boldsymbol{v}_j(\zeta)$ are analytic throughout the complex ζ-plane.

The solution of the linear equation (7.1.2) (or equivalently (7.1.14)) has been obtained by Caudrey [132] in terms of Jost functions $\boldsymbol{\phi}_j(X, \zeta)$ which have the asymptotic behaviour

$$\boldsymbol{\Phi}_j(X, \zeta) := \exp\{-\lambda_j(\zeta) X\} \boldsymbol{\phi}_j(X, \zeta) \to \boldsymbol{v}_j(\zeta) \text{ as } X \to -\infty. \tag{7.1.17}$$

Caudrey [132] showed how the Eq. (7.1.14) can be solved by expressing it as a Fredholm integral equation.

The complex ζ-plane is to be divided into regions such that, in the interior of each region, the order of the numbers $\mathrm{Re}(\lambda_i(\zeta))$ is fixed. As we pass from one region to another this order changes and hence, on a boundary between two regions, $\mathrm{Re}(\lambda_i(\zeta)) = \mathrm{Re}(\lambda_j(\zeta))$ for at least one pair $i \neq j$. The Jost function $\boldsymbol{\phi}_j$ is regular throughout the complex ζ-plane apart from poles and finite singularities on the boundaries between the regions. At any point in the interior of any region of the complex ζ-plane, the solution of Eq. (7.1.14) is obtained by the relation (2.12) from [132]. It is the direct spectral problem.

7.1.3 The spectral data

The information about the singularities of the Jost functions $\boldsymbol{\phi}_j(X, \zeta)$ reside in the spectral data. First let us consider the poles. It is assumed that a pole $\zeta_i^{(k)}$ in $\boldsymbol{\phi}_i(X, \zeta)$ is simple, does not coincide with a pole of $\boldsymbol{\phi}_j(X, \zeta)$, $j \neq i$, and does not lie on a boundary between two regions. Note that, for $\boldsymbol{\phi}_j(X, \zeta_i^{(k)})$, the point $\zeta_i^{(k)}$ lies in the interior of a regular region. First, we need the well-known relations for simple poles [132, 140]. Second, in Chapter 9 we will take into account the two-multiple poles. The residue of a simple pole can be calculated as [132, 140]

$$\operatorname{Res} \boldsymbol{\phi}_i(X, \zeta_i^{(k)}) = \sum_{\substack{j=1 \\ j \neq i}}^{n} \gamma_{ij}^{(k)} \boldsymbol{\phi}_j(X, \zeta_i^{(k)}) \tag{7.1.18}$$

and it can be found once we know the solution of (7.1.2) in any regular regions from solving the direct problem (see Section 7.1.2).

For convenience now we repeat the proof of relation (7.1.18) presented by Caudrey [132, 140]. If the Wronskian of fundamental solutions for the spectral equation

$$Wr = \det\left[\, \boldsymbol{\phi}_1(\lambda_1),\, \boldsymbol{\phi}_2(\lambda_2),\, \ldots,\, \boldsymbol{\phi}_n(\lambda_n) \,\right] \tag{7.1.19}$$

is non-zero at least at one point X_0, then it is proved in [141] (see p. 132 there) to be finite and non-zero even when ζ approaches a pole. Let $\boldsymbol{\phi}_1(\lambda_1(\zeta))$ have pole at $\zeta = \zeta_1^{(k)}$. Then $(\zeta - \zeta_1^{(k)})Wr = \det\left[\,(\zeta - \zeta_1^{(k)})\boldsymbol{\phi}_1(\lambda_1),\, \boldsymbol{\phi}_2(\lambda_2),\, \ldots,\, \boldsymbol{\phi}_n(\lambda_n)\,\right]$ and taking the limit $\zeta \to \zeta_1^{(k)}$ we obtain

$$0 = \det\left[\, \operatorname{Res} \boldsymbol{\phi}_1(\lambda_1),\, \boldsymbol{\phi}_2(\lambda_2),\, \ldots,\, \boldsymbol{\phi}_6(\lambda_6) \,\right]. \tag{7.1.20}$$

Thus the columns (vectors) are linearly dependent. Hence, the relation (7.1.18) has been proved.

The quantities $\zeta_i^{(k)}$ and $\gamma_{ij}^{(k)}$ constitute the discrete part of the spectral data.

Now we consider the singularities on the boundaries between regions. However, in order to simplify matters, we first make some observations. The solution of the spectral problem can be facilitated by using various symmetry properties. In view of (7.1.2), we need only consider the first elements of

$$\boldsymbol{\phi}_i(X, \zeta) = \begin{pmatrix} \phi_i(X, \zeta) \\ \phi_i(X, \zeta)_X \\ \phi_i(X, \zeta)_{XX} \end{pmatrix}, \tag{7.1.21}$$

while the symmetry

$$\phi_1(X, \zeta/\omega_1) = \phi_2(X, \zeta/\omega_2) = \phi_3(X, \zeta/\omega_3) \tag{7.1.22}$$

means we need only consider $\phi_1(X,\zeta)$. In our case, for $\phi_1(X,\zeta)$, the complex ζ-plane is divided into four regions by two lines (see Fig. 7.1) given by

$$\begin{aligned} &\text{(i)} \quad \zeta' = \omega_2\xi, \quad \text{where} \quad \mathrm{Re}(\lambda_1(\zeta)) = \mathrm{Re}(\lambda_2(\zeta)), \\ &\text{(ii)} \quad \zeta' = -\omega_3\xi, \quad \text{where} \quad \mathrm{Re}(\lambda_1(\zeta)) = \mathrm{Re}(\lambda_3(\zeta)), \end{aligned} \tag{7.1.23}$$

where ξ is real (see Fig. 7.1). The singularity of $\phi_1(X,\zeta)$ can appear only on these boundaries between the regular regions on the ζ-plane and it is characterized by functions $Q_{1j}(\zeta')$ at each fixed $j \neq 1$. We denote the limit of a quantity, as the boundary is approached, by the superfix $\pm$ in according to the sign of $\mathrm{Re}(\lambda_1(\zeta) - \lambda_j(\zeta))$ (see Fig. 7.1).

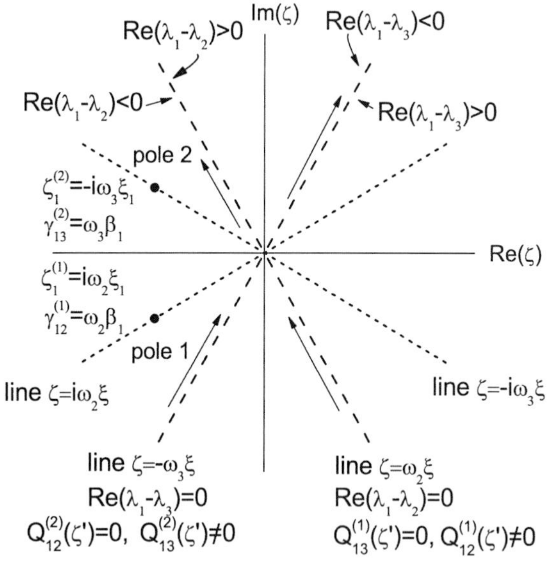

Figure 7.1: The regular regions for Jost functions $\phi_1(X,\zeta)$ in the complex ζ-plane. The dashed lines determine the boundaries between regular regions. These lines are lines where the singularity functions $Q_{1j}(\zeta')$ are given. The dotted lines are the lines where the poles appear.

In [132] (see Eq. (3.14) there) the jump of $\phi_1(X,\zeta)$ on the boundaries is calculated as

$$\phi_1^+(X,\zeta) - \phi_1^-(X,\zeta) = \sum_{j=2}^{3} Q_{1j}(\zeta)\phi_j^-(X,\zeta), \tag{7.1.24}$$

where, from (7.1.23), the sum is over the lines $\zeta' = \omega_2\xi$ and $\zeta' = -\omega_3\xi$ given by

$$\begin{aligned} &\text{(i)} \quad \zeta' = \omega_2\xi, \quad \text{with} \quad Q_{12}^{(1)}(\zeta') \neq 0, \quad Q_{13}^{(1)}(\zeta') \equiv 0, \\ &\text{(ii)} \quad \zeta' = -\omega_3\xi, \quad \text{with} \quad Q_{12}^{(2)}(\zeta') \equiv 0, \quad Q_{13}^{(2)}(\zeta') \neq 0. \end{aligned} \tag{7.1.25}$$

The singularity functions $Q_{1j}(\zeta')$ are determined by $W(X,0)$ through the matrix $\mathbf{B}(X,\zeta)$ (7.1.15) (see Eq. (3.13) in [132])

$$Q_{1j}(\zeta) = \frac{1}{\tilde{\boldsymbol{v}}_j(\zeta)\cdot\boldsymbol{v}_j(\zeta)}\tilde{\boldsymbol{v}}_j(\zeta)\cdot \int_{-\infty}^{\infty} \exp[(\lambda_1(\zeta)-\lambda_j(\zeta))z]\mathbf{B}(z,\zeta)\cdot\boldsymbol{\phi}_1^-(X,\zeta)\,dz. \tag{7.1.26}$$

The quantities $Q_{1j}(\zeta')$ along all the boundaries constitute the continuum part of the spectral data.

Thus, the spectral data are

$$S = \{\zeta_1^{(k)},\ \gamma_{1j}^{(k)},\ Q_{1j}(\zeta');\, j = 2,\ 3,\ k = 1,\ 2,\ \ldots,\ m\}. \tag{7.1.27}$$

One of the important features which is to be noted for the IST method is as follows. After the spectral data have been found from $\mathbf{B}(X,0;\zeta)$, i.e. at initial time, we need to seek the time-evolution of the spectral data from the equation (7.1.3). Analyzing (7.1.3) at $X\to\infty$ together with (7.1.17)

$$\phi_i(X,T,\zeta) = \exp\left[-\left(3\lambda_i(\zeta)\right)^{-1}T\right]\phi_i(X,0,\zeta),$$

the T-dependence is revealed as

$$\begin{aligned} \zeta_j^{(k)}(T) &= \zeta_j^{(k)}(0), \\ \gamma_{1j}^{(k)}(T) &= \gamma_{1j}^{(k)}(0)\exp\left\{\left[-\left(3\lambda_j(\zeta_1^{(k)})\right)^{-1} + \left(3\lambda_1(\zeta_1^{(k)})\right)^{-1}\right]T\right\}, \\ Q_{1j}(T;\zeta') &= Q_{1j}(0;\zeta')\exp\left\{\left[-\left(3\lambda_j(\zeta')\right)^{-1} + \left(3\lambda_1(\zeta')\right)^{-1}\right]T\right\}. \end{aligned} \tag{7.1.28}$$

The final step in the application of the IST method is to reconstruct $B(X,T;\zeta)$ from the evaluated spectral data. In the next section, we show how to do this.

7.1.4 The inverse spectral problem

The final procedure in IST method is that of the reconstruction of the matrix $B(X,T;\zeta)$ and $W(X,T)$ from the spectral data S.

The spectral data define $\Phi_1(X,\zeta)$ uniquely in the form (see Eq. (6.20) in [132]))

$$\begin{aligned}
&\Phi_1(X,T;\zeta) = \\
&1 - \sum_{k=1}^{K}\sum_{j=2}^{3} \gamma_{1j}^{(k)}(T)\frac{\exp\{[\lambda_j(\zeta_1^{(k)}) - \lambda_1(\zeta_1^{(k)})]X\}}{\lambda_1(\zeta_1^{(k)}) - \lambda_1(\zeta)}\Phi_1(X,T;\omega_j\zeta_1^{(k)}) \\
&+\frac{1}{2\pi\mathrm{i}}\int\sum_{j=2}^{3} Q_{1j}(T;\zeta')\frac{\exp\{[\lambda_j(\zeta') - \lambda_1(\zeta')]X\}}{\zeta' - \zeta}\Phi_1^-(X,T;\omega_j\zeta')d\zeta'.
\end{aligned} \tag{7.1.29}$$

Equation (7.1.29) contains the spectral data, namely K poles with the quantities $\gamma_{1j}^{(k)}$ for the bound state spectrum as well as the functions $Q_{1j}(\zeta')$ given along all the boundaries of regular regions for the continuous spectrum. The integral in (7.1.29) is along all the boundaries (see the dashed lines in Fig. 7.1). The direction of integration is taken so that the side chosen to be $\mathrm{Re}(\lambda_1(\zeta) - \lambda_j(\zeta)) < 0$ is shown by the arrows in Fig. 7.1 (for the lines (7.1.23), ξ sweeps from $-\infty$ to $+\infty$).

It is necessary to note that we should carry out the integration along the lines $\omega_2(\xi + \mathrm{i}\varepsilon)$ and $-\omega_3(\xi + \mathrm{i}\varepsilon)$ with $\varepsilon > 0$. In this case the condition (7.1.17) is satisfied. Passing to the limit $\varepsilon \to 0$ we can obtain the solution which does not satisfy the condition (7.1.17) (see Section 7.1.2). However, for any finite $\varepsilon > 0$, the restricted region on X can be determined where the solution associated with a finite $\varepsilon > 0$ (for which the condition (7.1.17) is valid) and the solution associated with $\varepsilon = 0$ are sufficiently close to each other. In this sense, taking the integration at $\varepsilon = 0$, we remain within the inverse scattering theory [132], and so the condition (7.1.17) can be omitted. The solution obtained at $\varepsilon = 0$ can be extended to sufficiently large finite X. Thus, we will interpret the solution obtained at $\varepsilon = 0$ as the solution of the VPE (5.1.10) which is valid for arbitrary but finite X.

By choosing appropriate values for ζ, the left-hand side in (7.1.29) can be $\Phi_1(X,T;\omega_j\zeta_1^{(k)})$, or by allowing ζ to approach the boundaries from the appropriate sides, the left-hand side can be $\Phi_1^-(X,T;\omega_j\zeta')$. We acquire a set of linear matrix/Fredholm equations in the unknowns $\Phi_1(X,T;\omega_j\zeta_1^{(k)})$ and $\Phi_1^-(X,T;\omega_j\zeta')$. The solution of this equation system enables one to define $\Phi_1(X,T;\zeta)$ from (7.1.29).

By knowing $\Phi_1(X,T;\zeta)$, we can take extra information into account, namely that the expansion of $\Phi_1(X,T;\zeta)$ as an asymptotic series in $\lambda_1^{-1}(\zeta)$ connects with $W(X,T)$ as follows (cf. Eq. (2.7) in [131]):

$$\Phi_1(X,T;\zeta) = 1 - \frac{1}{3\lambda_1(\zeta)}\,[W(X,T) - W(-\infty)] + O(\lambda_1^{-2}(\zeta)). \quad (7.1.30)$$

Consequently, the solution $W(X,T)$ and the matrix $B(X,T;\zeta)$ can be reconstructed from the spectral data.

In Sections 7.2 and 7.3, we show how the IST method can be applied to the VPE and the GVPE.

7.2 The multi-soliton solutions of the VPE by the inverse scattering method

In this Section the procedure for finding the exact N-soliton solution of the VPE via the inverse scattering method will be described [84, 91, 92, 93]. To do this we consider (7.1.29) with $Q_{1j}(\zeta) \equiv 0$. Then there is only the bound state spectrum which is associated with the soliton solutions.

Let the bound state spectrum be defined by K poles. The relation (7.1.29) is reduced to the form

$$\Phi_1(X,T;\zeta) = \quad (7.2.1)$$

$$= 1 - \sum_{k=1}^{K}\sum_{j=2}^{3} \gamma_{1j}^{(k)}(T)\frac{\exp\{[\lambda_j(\zeta_1^{(k)}) - \lambda_1(\zeta_1^{(k)})]X\}}{\lambda_1(\zeta_1^{(k)}) - \lambda_1(\zeta)}\Phi_1(X,T;\omega_j\zeta_1^{(k)}).$$

Eq. (7.2.1) involves the spectral data, namely the poles $\zeta_1^{(k)}$ and the quantities $\gamma_{1j}^{(k)}$. Until (7.2.13) we omit the T-dependence. First we will prove that $\operatorname{Re}\lambda = 0$ for compact support. From Eq. (7.1.2) we have

$$(\psi_X)_{XXX} + (U\psi_X)_X - \lambda\psi_X = 0, \quad (7.2.2)$$

and together with Eq. (7.1.2) this enables us to write

$$\frac{\partial}{\partial X}\left(\frac{\partial^2}{\partial X^2}\psi_X\psi^* - 3\psi_{XX}\psi_X^* + U\psi_X\psi^*\right) - 2\mathrm{Re}\,\lambda\,\psi_X\psi^* = 0. \quad (7.2.3)$$

Integrating Eq. (7.2.3) over all values of X, we obtain that, for compact support, $\mathrm{Re}\,\lambda = 0$ since, in the general case, $\int_{-\infty}^{\infty}\psi_X\psi^* dX \neq 0$.

As follows from Eqs. (2.12), (2.13), (2.36) and (2.37) of [131], $\psi_X(\zeta)$ is related to the adjoint states $\psi_X^A(-\zeta)$. In the usual manner, using the adjoint states and Eq. (14) from [133], and Eq. (2.37) from [131], one can obtain

$$\phi_{1\,X}(X,\zeta) = \quad (7.2.4)$$
$$= \frac{\mathrm{i}}{\sqrt{3}}\left[\phi_{1\,X}(X,-\omega_2\zeta)\phi_1(X,-\omega_3\zeta) - \phi_{1\,X}(X,-\omega_3\zeta)\phi_1(X,-\omega_2\zeta)\right].$$

It is easily seen that if $\zeta_1^{(1)}$ is a pole of $\phi_1(X,\zeta)$, then there is a pole either at $\zeta_1^{(2)} = -\omega_2\zeta_1^{(1)}$ (if $\phi_1(X,-\omega_2\zeta)$ has a pole), or at $\zeta_1^{(2)} = -\omega_3\zeta_1^{(1)}$ (if $\phi_1(X,-\omega_3\zeta)$ has a pole). For definiteness let $\zeta_1^{(2)} = -\omega_2\zeta_1^{(1)}$. Then, as follows from (7.2.4), $-\omega_3\zeta_1^{(2)}$ should be a pole. However, this pole coincides with the pole $\zeta_1^{(1)}$, since $-\omega_3\zeta_1^{(2)} = -\omega_3(-\omega_2)\zeta_1^{(1)} = \zeta_1^{(1)}$. Hence the poles appear in pairs, $\zeta_1^{(2n-1)}$ and $\zeta_1^{(2n)}$, under the condition $\zeta_1^{(2n)}/\zeta_1^{(2n-1)} = -\omega_2$, where n is the pair number.

Let us consider N pairs of poles, i.e. in all there are $K = 2N$ poles over which the sum is taken in (7.2.4). For the pair n ($n = 1,\, 2, \ldots,\, N$) we have the properties

$$\text{(i)}\quad \zeta_1^{(2n-1)} = \mathrm{i}\omega_2\xi_n, \qquad \text{(ii)}\quad \zeta_1^{(2n)} = -\mathrm{i}\omega_3\xi_n. \quad (7.2.5)$$

Since U is real and λ is imaginary, ξ_k is real. The relationships (7.2.5) are in line with the condition (2.33) from [131]. These relationships are also similar to Eqs. (6.24) and (6.25) in [132], while $\gamma_{1j}^{(k)}$ turns out to be different from $\tilde{\gamma}_{1j}^{(k)}$ for the Boussinesq equation (see Eqs. (6.24) and (6.25) in [132]). Indeed, by considering (7.2.4) in the vicinity of the first pole $\zeta_1^{(2n-1)}$ of the pair n and using the relation (7.2.1), one can obtain a relation between $\gamma_{12}^{(k)}$ and $\gamma_{13}^{(k)}$. In this case the functions $\phi_{1,X}(X,\zeta)$, $\phi_1(X,-\omega_2\zeta)$, $\phi_{1,X}(X,-\omega_2\zeta)$ also have poles here, while the functions $\phi_1(X,-\omega_3\zeta)$, $\phi_{1,X}(X,-\omega_3\zeta)$ do not have poles here. Substituting $\phi_1(X, \zeta)$ in the form (7.2.1) into Eq. (7.2.4) and letting $X \to -\infty$, we have the ratio $\gamma_{13}^{(2n)}/\gamma_{12}^{(2n-1)} = \omega_2$ and $\gamma_{12}^{(2n)} = \gamma_{13}^{(2n-1)} = 0$. Therefore the properties of $\gamma_{ij}^{(k)}$ should be defined by the relationships

$$\begin{array}{llll} \text{(i)} & \gamma_{12}^{(2n-1)} = \omega_2 \beta_k, & & \gamma_{13}^{(2n-1)} = 0, \\ \text{(ii)} & \gamma_{12}^{(2n)} = 0, & & \gamma_{13}^{(2n)} = \omega_3 \beta_k, \end{array} \tag{7.2.6}$$

where, as it will be proved below, β_k is real when $U = W_X$ is real.

The equation (7.2.1) allows us to define the functions $\Phi_1(X, \zeta)$. Indeed, substituting the values $\zeta = \omega_1 \zeta_1^{(k)}$, $\zeta = \omega_3 \zeta_2^{(k)}$ in the left-hand side of these equations, we obtain a system of $2N$ linear algebraic equations in the unknowns $\Phi_1(X, \omega_2 \zeta_1^{(k)})$, $\Phi_1(X, \omega_3 \zeta_1^{(k)})$. Hence, we could take the function $\Phi_1(X, \zeta)$ from Eq. (7.2.1).

However, there is a more direct method, in which there is no need to obtain the variables $\Phi_1(X, \omega_2 \zeta_1^{(k)})$, $\Phi_1(X, \omega_3 \zeta_1^{(k)})$ explicitly. It turns out that we need to calculate only a determinant of some matrix. This approach is similar to the method referred to in [132, 140, 84, 85, 86, 87]. It is convenient to use new variables introduced by the definition

$$\Psi_k(X) = \sum_{j=2}^{3} \gamma_{1j}^{(k)}(0) \exp\{\lambda_j(\zeta_1^{(k)})X\} \Phi_1(X; \omega_j \zeta_1^{(k)}). \tag{7.2.7}$$

We may rewrite the relationship (7.2.1) as

$$\Phi_1(X; \zeta) = 1 - \sum_{k=1}^{2N} \frac{\exp\{-\lambda_1(\zeta_1^{(k)})X\}}{\lambda_1(\zeta_1^{(k)}) - \lambda_1(\zeta)} \Psi_k(X). \tag{7.2.8}$$

Taking into account (7.1.30), namely

$$\Phi_1(X; \zeta) = 1 - \frac{1}{3\lambda_1(\zeta)} [W(X) - W(-\infty)] + O(\lambda_1^{-2}(\zeta)),$$

and (7.2.7) and (7.2.8), the following relationship may be found

$$-\frac{1}{3} [W(X) - W(-\infty)] = \sum_{k=1}^{2N} \exp\{-\lambda_1(\zeta_1^{(k)})X\} \Psi_k(X). \tag{7.2.9}$$

Eq. (7.2.8) with (7.2.7) can be rewritten as follows:

$$\exp(\lambda_1(\zeta)X)\Phi_1(X;\zeta) = \exp(\lambda_1(\zeta)X) - \sum_{k=1}^{2N} \frac{\exp\{(\lambda_1(\zeta)-\lambda_1(\zeta_1^{(k)}))X\}}{\lambda_1(\zeta_1^{(k)})-\lambda_1(\zeta)}\Psi_k(X). \tag{7.2.10}$$

Substituting the values $\zeta=\omega_2\zeta_1^{(k)}$, $\zeta=\omega_3\zeta_1^{(k)}$ in the left-hand side of these equations, we obtain a system of $2N$ linear algebraic equations in the unknowns $\Psi_k(X)$, for $k=1,\ldots,2N$. The matrix form of this system of equations is

$$\mathbf{M}\boldsymbol{\Psi} = \mathbf{b}, \tag{7.2.11}$$

where

$$\boldsymbol{\Psi} = \begin{pmatrix} \Psi_1(X) \\ \Psi_2(X) \\ \cdots \\ \Psi_{2N}(X) \end{pmatrix}, \quad \mathbf{b} = \begin{pmatrix} \gamma_{12}^{(1)}\exp(\omega_2\zeta_1^{(1)}X) \\ \gamma_{13}^{(2)}\exp(\omega_3\zeta_1^{(1)}X) \\ \cdots \\ \gamma_{13}^{(2N)}\exp(\omega_3\zeta_1^{(N)}X) \end{pmatrix}. \tag{7.2.12}$$

With account of the T-evolution (7.1.2) for the spectral data, the elements of $2N\times 2N$ matrix $\mathbf{M}$ are

$$M_{kl}(X,T) = \delta_{kl} - \tag{7.2.13}$$

$$-\sum_{j=2}^{3}\gamma_{1j}^{(k)}(0)\frac{\exp\left\{\left[-\left(3\lambda_j(\zeta_1^{(k)})\right)^{-1}+\left(3\lambda_1(\zeta_1^{(k)})\right)^{-1}\right]T+\left(\lambda_j(\zeta_1^{(k)})-\lambda_1(\zeta_1^{(l)})\right)X\right\}}{\lambda_j(\zeta_1^{(k)})-\lambda_1(\zeta_1^{(l)})},$$

and

$$n=1,\ 2,\ldots,\ N,$$
$$\lambda_1(\zeta_1^{(2n-1)}) = i\omega_2\xi_n, \qquad \lambda_2(\zeta_1^{(2n-1)}) = i\omega_3\xi_n,$$

$$\lambda_1(\zeta_1^{(2n)}) = -i\omega_3\xi_n, \qquad \lambda_3(\zeta_1^{(2n)}) = -i\omega_2\xi_n,$$

$$\gamma_{12}^{(2n-1)} = \omega_2\beta_n, \qquad \gamma_{13}^{(2n-1)} = 0,$$

$$\gamma_{12}^{(2n)} = 0, \qquad \gamma_{13}^{(2n)} = \omega_3\beta_n.$$

Since for any column j of the matrix $\mathbf{M}$ we have

$$\exp(\omega_l \xi_n X)\frac{\partial}{\partial X}M_{ij} = b_i, \quad l = \begin{cases} 2, & \text{if} \quad i = 2n-1 \\ 3, & \text{if} \quad i = 2n \end{cases},$$

the sum for (7.2.9) is

$$\sum_{k=1}^{2N} \exp\{-\lambda_1(\zeta_1^{(k)})X\}\Psi_k(X,T) = \frac{1}{\det \mathbf{M}}\frac{\partial \det \mathbf{M}}{\partial X}.$$

Consequently, from the relation (7.2.9), the following key relationship may be obtained

$$W(X,T) - W(-\infty) = 3\frac{\partial}{\partial X}\ln(\det \mathbf{M}(X,T)). \tag{7.2.14}$$

For the N-soliton solution there are N arbitrary constants ξ_n and N arbitrary constants β_n.

The final result for the N-soliton solution of the VPE is defined by relationship (7.2.14) with (7.2.13).

7.2.1 Examples of one- and two-soliton solutions

In order to obtain the one-soliton solution of the VPE (5.1.10)

$$W_{XXT} + (1 + W_T)W_X = 0,$$

we need first to calculate the 2×2 matrix $\mathbf{M}$ according to (7.2.13) with $N = 1$. We find that the matrix is

$$\begin{pmatrix} \begin{array}{c} 1 - \dfrac{\omega_2\beta_1}{\sqrt{3}\xi_1}\times \\ \times\exp[\sqrt{3}\xi_1 X - (\sqrt{3}\xi_1)^{-1}T] \end{array} & \begin{array}{c} \dfrac{\mathrm{i}\omega_3\beta_1}{2\xi_1}\times \\ \times\exp[2\mathrm{i}\omega_3\xi_1 X - (\sqrt{3}\xi_1)^{-1}T] \end{array} \\ \\ \begin{array}{c} \dfrac{-\mathrm{i}\omega_2\beta_1}{2\xi_1}\times \\ \times\exp[-2\mathrm{i}\omega_2\xi_1 X - (\sqrt{3}\xi_1)^{-1}T] \end{array} & \begin{array}{c} 1 - \dfrac{\omega_3\beta_1}{\sqrt{3}\xi_1}\times \\ \times\exp[\sqrt{3}\xi_1 X - (\sqrt{3}\xi_1)^{-1}T] \end{array} \end{pmatrix} \tag{7.2.15}$$

and its determinant is

$$\det \mathbf{M}(X,T) = \left\{1 + \frac{\beta_1}{2\sqrt{3}\xi_1}\exp\left[\sqrt{3}\xi_1\left(X - \frac{T}{3\xi_1^2}\right)\right]\right\}^2. \tag{7.2.16}$$

Consequently, from Eq. (7.2.14) we have the one-soliton solution of the VPE

$$U(X,T)=W_X(X,T)=\frac{9}{2}\xi_1^2\,\mathrm{sech}^2\left[\frac{\sqrt{3}}{2}\xi_1\left(X-\frac{T}{3\xi_1^2}\right)+\alpha_1\right], \tag{7.2.17}$$

where $\alpha_1=\frac{1}{2}\ln(\beta_1/2\sqrt{3}\xi_1)$ is an arbitrary constant. Since U is real, it follows from (7.2.17) that β_1 is real. By writing $\sqrt{3}\xi_1/2=k$ in (7.2.17), with the condition $\beta_1/\xi_1>0$ we recover the one-soliton solution as we found previously by Hirota's method (see Eq. (3.4) in [74] and/or (6.5.14)).

Note that with $\beta_1/\xi_1<0$ we have the real solution in the form of the singular soliton (**??**) [123]. Analysis of the singular soliton solution is presented in Section 8.4.

It is of interest to compare Eq. (7.2.17) with the solution of the fifth-order KdV-like equation discussed in [131]. The spectral equation (7.1.2) is the same as that given by (1.1) (with $R=0$) in [131], whereas the equation that governs the time dependence of ψ, i.e. (7.1.3), is different from (1.2) in [131]. Thus the X dependence of (7.2.16) should agree with the x dependence of the solution given by (3.30) in [131]. With the identification $U=6Q$, $\xi_1=\eta$, this is indeed the case.

Let us now consider the two-soliton solution of the VPE. In this case $\mathbf{M}(X,T)$ is a 4×4 matrix. We will not give the explicit form here, but we find that

$$\det\mathbf{M}(X,T)=\left(1+q_1^2+q_2^2+b^2q_1^2q_2^2\right)^2, \tag{7.2.18}$$

where

$$q_i=\exp\left[\frac{\sqrt{3}}{2}\xi_i\left(X-\frac{T}{3\xi_i^2}\right)+\alpha_i\right],\quad b^2=\left(\frac{\xi_2-\xi_1}{\xi_2+\xi_1}\right)^2\frac{\xi_1^2+\xi_2^2-\xi_1\xi_2}{\xi_1^2+\xi_2^2+\xi_1\xi_2}, \tag{7.2.19}$$

and $\alpha_i=\frac{1}{2}\ln(\beta_i/2\sqrt{3}\xi_i)$ are arbitrary constants. The two-soliton solution to the VPE as found by the IST method is given by (7.2.9) together with (7.2.18). With the identification $\sqrt{3}\xi_i/2=k_i$ ($i=1$, 2) we recover the two-soliton solution as given by Hirota's method (see Eqs. (4.1)–(4.5) in [74] and/or (6.5.24)).

Finally we note that comparison of (7.2.9) with $W = 6(\ln f)_X$ from (6.2.2) shows that

$$\ln(\det \mathbf{M}(X,T)) = 2\ln(f). \tag{7.2.20}$$

so that $\det \mathbf{M}(X,T)$ is a perfect square for arbitrary N.

7.3 The inverse scattering transform method for the generalised VPE

In this Section, following [90], we extend the investigation of the VPE by means of the IST method as given in Section 7.2 (and [84]) to the generalised Vakhnenko-Parkes equation (GVPE) [101].

7.3.1 The bilinear Hirota form of the GVPE

The transformed version of the GVE (4.5.5)

$$\frac{\partial}{\partial x}\left(\mathcal{D}^2 u + \tfrac{1}{2}u^2 + \beta u\right) + \mathcal{D}u = 0 \tag{7.3.1}$$

has the form given by (3.5) in [101], namely

$$U_{XXT} + UU_T + U_X\int_{-\infty}^{X} U_T(X',T)\,dX' + U_X + \beta U_T = 0, \tag{7.3.2}$$

or equivalently

$$W_{XXT} + (1+W_T)W_X + \beta W_T = 0. \tag{7.3.3}$$

in variables

$$x = \theta(X,T) := T + W(X,T) + x_0, \tag{7.3.4}$$

$$t = X, \quad u(x,t) = U(X,T) = W_X(X,T),$$

where β is a real arbitrary constant.

Since the VE as transformed by means of (7.3.4) is known as the VPE, we refer to the transformed GPE (7.3.2) or (7.3.3), as obtained by means of the same transformation (7.3.4), as the generalised Vakhnenko-Parkes equation (GVPE). The GVPE was solved by the Hirota method in Section 6.6.

In bilinear Hirota form the equation (7.3.3) is

$$F(D_X, D_T)f \cdot f = 0, F(D_X, D_T) := D_X^3 D_T + D_X^2 + \beta D_X D_T. \quad (7.3.5)$$

We noted in Section 4.5 that with $\beta = -1$ and $T \to -T$, (7.3.3) and (7.3.5) are associated with the Hirota-Satsuma equation (HSE) for shallow water waves [35, 100]. As far as we are aware, the solution by the IST method to the HSE (i.e. equation (7.3.2) with $\beta = -1$ and $T \to -T$) has not been given explicitly in the literature. In Sections 7.3.2 and 7.3.3, following [90], we present the IST method to solve (7.3.2) for arbitrary non-zero β and hence find the N-soliton solution to (7.3.1) subject to the boundary condition $u \to 0$ as $|x| \to \infty$.

7.3.2 Bäcklund transformation, Lax pair and conservation laws for the GVPE

We will show that the Bäcklund transformation for (7.3.5) is given by the two equations

$$(D_X^3 + \beta D_X - \lambda(X))f' \cdot f = 0, \quad (7.3.6)$$
$$(3D_X D_T + 1 + \mu(T) D_X)f' \cdot f = 0, \quad (7.3.7)$$

where $\lambda(X)$ is an arbitrary function of X and $\mu(T)$ is an arbitrary function of T. We follow the method developed in [114].

Consider the expression P defined by

$$P := [(D_T D_X^3 + D_X^2 + \beta D_X D_T)f' \cdot f']ff - \quad (7.3.8)$$
$$-f'f'[(D_T D_X^3 + D_X^2 + \beta D_X D_T)f \cdot f],$$

where $f \neq f'$. In [84] it was shown that

$$[(D_T D_X^3 + D_X^2)f' \cdot f']ff - f'f'[(D_T D_X^3 + D_X^2)f \cdot f] \quad (7.3.9)$$
$$= 2D_T(D_X^3 f' \cdot f) \cdot (f'f) - 2D_X(\{3D_T D_X + 1\}f' \cdot f) \cdot (D_X f' \cdot f).$$

By using (7.3.9) and the identities (II.1) and (VII.2) from [96], P given by (7.3.8) can be reduced to the form

$$P = 2D_T(\{D_X^3 + \beta D_X - \lambda(X)\}f' \cdot f) \cdot (f'f) -$$
$$-2D_X(\{3D_T D_X + 1 + \mu(T) D_X\}f' \cdot f) \cdot (D_X f' \cdot f). \quad (7.3.10)$$

It is clear from (7.3.10) that if (7.3.6) and (7.3.7) hold then $P = 0$. Furthermore it then follows from (7.3.8) that if f is a solution of (7.3.5) then so is f' and vice-versa. Consequently, we have proved that the two equations (7.3.6) and (7.3.7) constitute a Bäcklund transformation for equation (7.3.5). As expected, with $\beta = -1$ and $T \to -T$, (7.3.6) and (7.3.7) become the Bäcklund transformation for the HSE (see equations (5.131) and (5.132) in [35]).

The inclusion of μ in the operator $3D_T + \mu(T)$ which appears in (7.3.7) corresponds to a multiplication of f and f' by terms of the form $e^{g(T)}$ and $e^{g'(T)}$ respectively; from $W = 6(\ln f)_X$ (6.2.2) we see that this has no effect on W or W'. Hence, without loss of generality, we may take $\mu = 0$ in (7.3.7) if we wish.

By introducing the function

$$\psi = f'/f, \tag{7.3.11}$$

and taking into account (6.2.2), we find that (7.3.6) and (7.3.7) reduce to

$$\psi_{XXX} + (\beta + W_X)\psi_X - \lambda\psi = 0, \tag{7.3.12}$$

$$3\psi_{XT} + (1 + W_T)\psi + \mu\psi_X = 0 \tag{7.3.13}$$

respectively, where we have used results similar to (X.1) – (X.3) in [35].

From (7.3.12) and (7.3.13) it can be shown that

$$3\lambda\psi_T + (1+W_T)\psi_{XX} - W_{XT}\psi_X + [W_{XXT} + (\beta+W_X)(1+W_T) + \mu\lambda]\psi = 0 \tag{7.3.14}$$

and

$$[W_{XXT} + (1 + W_T)W_X + \beta W_T]_X\psi + (3\psi_T + \mu\psi)\lambda_X = 0. \tag{7.3.15}$$

In view of (7.3.3), (7.3.13) becomes

$$3\lambda\psi_T + (1 + W_T)\psi_{XX} - W_{XT}\psi_X + (\beta + \lambda\mu)\psi = 0, \tag{7.3.16}$$

and (7.3.15) implies that $\lambda_X = 0$ so the spectrum λ of (7.3.12) remains constant. Constant λ is what is required in the IST problem discussed in Section 7.3.3. (7.3.12) and (7.3.16) are the Lax pair for (7.3.3). As expected, with $\beta = -1$ and $T \to -T$, (7.3.12), (7.3.13) and (7.3.16) are the corresponding equations for the HSE (cf. equations (A8a), (A8b) and (A10b) respectively in [136]).

Following the procedure given in [35, 115], we can rewrite (7.3.12) and (7.3.16) in terms of the potential W. Recalling that $\psi = f'/f$, and noting that $W' - W = 6\varphi_X$ and $W' + W = 6\rho_X$, where $\varphi = \ln f'/f$ and $\rho = \ln f'f$, we find that (7.3.12) and (7.3.16) give the following Bäcklund transformation in ordinary form:

$$\begin{aligned}(W'-W)_{XX} + \tfrac{1}{2}(W'-W)(W'+W)_X + \tfrac{1}{36}(W'-W)^3 \\ +\beta(W'-W) - 6\lambda = 0, \qquad (7.3.17)\end{aligned}$$

$$\begin{aligned}3\lambda(W'-W)_T + \left[(1+W_T)((W'-W)_X + \tfrac{1}{6}(W'-W)^2)\right. \\ \left. - W_{XT}(W'-W)\right]_X = 0. \qquad (7.3.18)\end{aligned}$$

A systematic way to derive higher conservation laws via the Bäcklund transformation has been developed by Satsuma; he applied it to the KdV equation [116]. Later Satsuma and Kaup [115] applied the method to a higher order KdV equation. Since our Eq. (7.3.17) is (apart from a scaling factor) the same as Eq. (29) in [115], and our Eq. (7.3.18) is in conservation form, we can apply the results on higher conservation laws in §4 of [115] to the GVPE in the form (7.3.2). Thus we deduce that an infinite sequence of conservation laws is associated with (7.3.2). For example, the first two nontrivial conserved densities are U and $(U^3 - 3U_X^2 + 3\beta U^2)$.

7.3.3 The IST problem and its N-soliton solution

As shown in Section 7.3.2, the IST problem for the GVPE (7.3.3) has a spectral equation for ψ of third order, namely (7.3.12). The inverse problem for certain third-order spectral equations has been considered by Kaup [131] and Caudrey [132, 133]. The time evolution of ψ is determined from (7.3.13) or (7.3.16).

Following the method described by Caudrey [132], the spectral equation (7.3.12) can be rewritten

$$\frac{\partial}{\partial X}\boldsymbol{\psi} = [\mathbf{A}(\zeta) + \mathbf{B}(X,\zeta)]\cdot\boldsymbol{\psi} \qquad (7.3.19)$$

with

$$\boldsymbol{\psi} = \begin{pmatrix} \psi \\ \psi_X \\ \psi_{XX} \end{pmatrix}, \qquad \mathbf{A} = \begin{pmatrix} 0 & 1 & 0 \\ 0 & 0 & 1 \\ \lambda & -\beta & 0 \end{pmatrix}, \qquad \mathbf{B} = \begin{pmatrix} 0 & 0 & 0 \\ 0 & 0 & 0 \\ 0 & -W_X & 0 \end{pmatrix}. \qquad (7.3.20)$$

We find the eigenvalue $\lambda_j(\zeta)$ of the matrix $\mathbf{A}$ from the equation

$$\det(\mathbf{A} - \lambda_j\mathbf{E}) = -\lambda_j^3 - \beta\lambda_j + \lambda = 0, \qquad (7.3.21)$$

where **E** is the identity matrix. The relation (7.3.21) between the values λ and λ_j can be rewritten in parametric form with ζ as parameter, namely

$$\lambda_j = \left(\frac{\beta}{3}\right)^{1/2} \left(\omega_j \zeta - \frac{1}{\omega_j \zeta}\right), \tag{7.3.22}$$

$$\lambda = \left(\frac{\beta}{3}\right)^{3/2} \left(\zeta^3 - \frac{1}{\zeta^3}\right), \tag{7.3.23}$$

where $\omega_j = e^{i2\pi(j-1)/3}$ are the cube of roots of 1 ($j = 1, 2, 3$). Because of the properties $\lambda_1(\zeta) = \lambda_1(-\zeta^{-1})$, $\lambda_2(\zeta) = \lambda_3(-\zeta^{-1})$' $\lambda_3(\zeta) = \lambda_2(-\zeta^{-1})$ and $\lambda(\zeta) = \lambda(-\zeta^{-1})$, it is sufficient to consider the values ζ located outside (or inside) of the circle $|\zeta| = 1$ only.

The right- and left-eigenvectors are

$$\boldsymbol{v}_j(\zeta) = \begin{pmatrix} 1 \\ \lambda_j \\ \lambda_j^2 \end{pmatrix}, \qquad \tilde{\boldsymbol{v}}_j(\zeta) = \begin{pmatrix} \lambda_j^2 + \beta, & \lambda_j, & 1 \end{pmatrix}. \tag{7.3.24}$$

It should be noted that the passage to the limit $\beta \to 0$ must be carried out with $\sqrt{\beta}\,\zeta$ held constant.

The general theory of the inverse scattering problem for N spectral equations has been developed in [132]. The solution of the linear equation (7.3.19) (or equivalently (7.3.3)) has been obtained by Caudrey [132] in terms of Jost functions $\boldsymbol{\phi}_j(X, \zeta)$ which have the asymptotic behaviour

$$\boldsymbol{\Phi}_j(X, \zeta) := \exp\{-\lambda_j(\zeta)X\}\boldsymbol{\phi}_j(X, \zeta) \to \boldsymbol{v}_j(\zeta) \text{ as } X \to -\infty. \tag{7.3.25}$$

Here T is regarded as a parameter until the T-evolution of the scattering data is taken into account later. The solution of the direct problem is given by the equation system (4.5) in [132]. We shall restrict our attention to the N-soliton solution. To do this we consider equation (6.20) from [132] by putting $Q_{ij}(\zeta) \equiv 0$. Then there is only the bound state spectrum which is associated with the soliton solutions.

Let the bound state spectrum be defined by K poles located, for definiteness, outside the circle $|\zeta| = 1$. The relation (4.5) from [132] is reduced to the form

$$\Phi_1(X, \zeta) = 1 - \sum_{k=1}^{K} \sum_{j=2}^{3} \gamma_{1j}^{(k)} \frac{\exp\{[\lambda_j(\zeta_1^{(k)}) - \lambda_1(\zeta_1^{(k)})]X\}}{\lambda_1(\zeta_1^{(k)}) - \lambda_1(\zeta)} \Phi_1(X, \omega_j \zeta_1^{(k)}).$$

(7.3.26)

We need only consider the function $\Phi_1(X,\zeta)$ since there is a set of symmetry properties as for the Boussinesq equation, namely the properties (6.15), (6.16) in [132] for Jost functions $\phi_j(X,\zeta)$

$$\phi_1(X,\zeta/\omega_1)=\phi_2(X,\zeta/\omega_2)=\phi_3(X,\zeta/\omega_3),\;\;\phi_j(X,\zeta)=\phi_j(X,-\zeta^{-1}) \tag{7.3.27}$$

that follow from equations (7.3.20) and (7.3.22).

(7.3.26) involves the spectral data, namely the poles $\zeta_i^{(k)}$ and the quantities $\gamma_{ij}^{(k)}$. First we will prove that $\operatorname{Re}\lambda=0$ for compact support. Indeed, from (7.3.12) we have

$$(\psi_X)_{XXX}+[(\beta+W_X)\psi_X]_X-\lambda\psi_X=0, \tag{7.3.28}$$

and together with (7.3.13) this enables us to write

$$\frac{\partial}{\partial X}\left(\frac{\partial^2}{\partial X^2}\psi_X\psi^*-3\psi_{XX}\psi_X^*+(\beta+W_X)\psi_X\psi^*\right)-2\operatorname{Re}\lambda\,\psi_X\psi^*=0. \tag{7.3.29}$$

Integrating (7.3.29) over all values of X, we obtain that for compact support $\operatorname{Re}\lambda=0$ since, in the general case, $\int_{-\infty}^{\infty}\psi_X\psi^* dX\neq 0$.

As follows from equations (2.12), (2.13), (2.36) and (2.37) of [131], $\psi_X(\zeta)$ is related to the adjoint states $\psi^A(-\zeta)$. In the usual manner, using the adjoint states and equation (14) from [133], and equation (2.37) from [131], one can obtain

$$\phi_{1\,X}(X,\zeta)=\frac{\mathrm{i}}{\sqrt{3}}[\phi_{1\,X}(X,-\omega_2\zeta)\phi_1(X,-\omega_3\zeta)- \tag{7.3.30}$$

$$-\phi_{1\,X}(X,-\omega_3\zeta)\phi_1(X,-\omega_2\zeta)].$$

It is easily seen that if $\zeta_1^{(1)}$ is a pole of $\phi_1(X,\zeta)$, then there is a pole either at $\zeta_1^{(2)}=-\omega_2\zeta_1^{(1)}$ (if $\phi_1(X,-\omega_2\zeta)$ has a pole), or at $\zeta_1^{(2)}=-\omega_3\zeta_1^{(1)}$ (if $\phi_1(X,-\omega_3\zeta)$ has a pole). For definiteness, let $\zeta_1^{(2)}=-\omega_2\zeta_1^{(1)}$, then as follows from equation (7.3.30) the point $-\omega_3\zeta_1^{(2)}$ should be a pole. However, this pole coincides with the pole $\zeta_1^{(1)}$, since $-\omega_3\zeta_1^{(2)}=-\omega_3(-\omega_2)\zeta_1^{(1)}=\zeta_1^{(1)}$. Hence, the poles appear in pairs $\zeta_1^{(2n-1)}$, $\zeta_1^{(2n)}$ under the condition $\zeta_1^{(2n)}/\zeta_1^{(2n-1)} = -\omega_2$, where n is the number pair.

Let us consider N pairs of poles, i.e. in all there are $K = 2N$ poles over which the sum is taken in (7.3.26). For the pair n ($n = 1, 2, \ldots, N$) we have the properties

$$\text{(i)} \quad \zeta_1^{(2n-1)} = \mathrm{i}\omega_2\xi_n, \qquad \text{(ii)} \quad \zeta_1^{(2n)} = -\mathrm{i}\omega_3\xi_n. \tag{7.3.31}$$

Since U is real and λ is imaginary, either ξ_n is real when $\beta > 0$ or ξ_n is imaginary when $\beta < 0$, i.e. $\sqrt{\beta}\,\xi_n$ is real.

By considering equation (7.3.30) in the vicinity of the first pole $\zeta_1^{(2n-1)}$ of the pair n and using the relation (7.3.26), one can obtain a relation between $\gamma_{12}^{(2n-1)}$ and $\gamma_{13}^{(2n)}$. In this case the functions $\phi_{1\,X}(X,\zeta)$, $\phi_1(X,-\omega_2\zeta)$, $\phi_{1\,X}(X,-\omega_2\zeta)$ also have poles here, while the functions $\phi_1(X,-\omega_3\zeta)$, $\phi_{1\,X}(X,-\omega_3\zeta)$ do not have poles here. Substituting $\phi_1(X,\zeta)$ in the form given by (7.3.25) and (7.3.26) into equation (7.3.30) and letting $X \to -\infty$, we have $\gamma_{12}^{(2n)} = \gamma_{13}^{(2n-1)} = 0$ and the ratio

$$\frac{\gamma_{12}^{(2n-1)}}{\gamma_{13}^{(2n)}} = \frac{\omega_2\xi_n + (\omega_2\xi_n)^{-1}}{\omega_3\xi_n + (\omega_3\xi_n)^{-1}}. \tag{7.3.32}$$

Therefore the properties of $\gamma_{ij}^{(k)}$ should be defined by the relationships

$$\left.\begin{aligned} &\text{(i)} \quad \gamma_{12}^{(2n-1)} = \sqrt{\beta}\,\gamma_n[\omega_2\xi_n + (\omega_2\xi_n)^{-1}], \quad \gamma_{13}^{(2n-1)} = 0, \\ &\text{(ii)} \quad \gamma_{12}^{(2n)} = 0, \qquad \gamma_{13}^{(2n)} = \sqrt{\beta}\,\gamma_n[\omega_3\xi_n + (\omega_3\xi_n)^{-1}], \end{aligned}\right\} \tag{7.3.33}$$

where γ_n are arbitrary constants. We will show below that γ_n is real when W_X is real.

Following [84] we expand $\Phi_1(X,\zeta)$ as an asymptotic series in $\lambda_1^{-1}(\zeta)$ to obtain

$$\Phi_1(X,\zeta) = 1 - \frac{1}{3\lambda_1(\zeta)}\,[W(X) - W(-\infty)] + O(\lambda_1^{-2}(\zeta)). \tag{7.3.34}$$

On the other hand, we may rewrite the relationship (7.3.26) as (see, for instance, equations (6.33) and (6.34) in [132])

$$\Phi_1(X,\zeta) = 1 - \sum_{k=1}^{K} \frac{\exp\{-\lambda_1(\zeta_1^{(k)})X\}}{\lambda_1(\zeta_1^{(k)}) - \lambda_1(\zeta)} \Psi_k(X), \tag{7.3.35}$$

$$\Psi_k(X) = \sum_{j=2}^{3} \gamma_{1j}^{(k)} \exp\{\lambda_j(\zeta_1^{(k)})X\}\Phi_1(X,\omega_j\zeta_1^{(k)}).$$

From (7.3.34) and (3.2.19) it may be shown that (cf. equation (6.38) in [132])

$$W(X)-W(-\infty) = -3\sum_{k=1}^{K} \exp\{-\lambda_1(\zeta_1^{(k)})X\}\Psi_k(X) = 3\frac{\partial}{\partial X}\ln(\det \mathbf{M}). \tag{7.3.36}$$

The matrix $\mathbf{M}$ is defined as in the relationship (6.36) in [132] by

$$M_{kl}(X) = \delta_{kl} - \sum_{j=2}^{3} \gamma_{1j}^{(k)} \frac{\exp\{[\lambda_j(\zeta_1^{(k)}) - \lambda_1(\zeta_1^{(l)})]X\}}{\lambda_j(\zeta_1^{(k)}) - \lambda_1(\zeta_1^{(l)})}. \tag{7.3.37}$$

Now let us consider the T-evolution of the spectral data. By analyzing the solution of equation (7.3.13) when $X \to -\infty$ together with (7.3.25), we find that

$$\phi_i(X,T,\zeta) = \exp\left[-\left(3\lambda_i(\zeta)\right)^{-1} T\right] \phi_i(X,0,\zeta).$$

Hence the T-evolution of the scattering data is given by the relationships (with $k = 1,\ 2, \dots,\ K$)

$$\begin{aligned} &\zeta_j^{(k)}(T) = \zeta_j^{(k)}(0), \\ &\gamma_{1j}^{(k)}(T) = \gamma_{1j}^{(k)}(0)\exp\left\{\left[-\left(3\lambda_j(\zeta_1^{(k)})\right)^{-1} + \left(3\lambda_1(\zeta_1^{(k)})\right)^{-1}\right]T\right\}. \end{aligned} \tag{7.3.38}$$

The final result, including the T-evolution, for the N-soliton solution of the GVPE (7.3.2) is

$$U(X,T) = W_X(X,T) = 3\frac{\partial^2}{\partial X^2}\ln\left(\det \mathbf{M}(X,T)\right), \tag{7.3.39}$$

where $\mathbf{M}$ is the $2N\times 2N$ matrix given by

$$M_{kl} = \delta_{kl} - \sum_{j=2}^{3} \gamma_{1j}^{(k)}(0) \times \\ \times \frac{\exp\left\{\left[-\left(3\lambda_j(\zeta_1^{(k)})\right)^{-1}+\left(3\lambda_1(\zeta_1^{(k)})\right)^{-1}\right]T+\left(\lambda_j(\zeta_1^{(k)})-\lambda_1(\zeta_1^{(l)})\right)X\right\}}{\lambda_j(\zeta_1^{(k)})-\lambda_1(\zeta_1^{(l)})}, \tag{7.3.40}$$

and

$$n = 1,\ 2, \ldots,\ N, \qquad m = 2n-1,$$

$$\lambda_1(\zeta_1^{(m)}) = \mathrm{i}\sqrt{\beta/3}\,[\omega_2\xi_m + (\omega_2\xi_m)^{-1}],$$

$$\lambda_2(\zeta_1^{(m)}) = \mathrm{i}\sqrt{\beta/3}\,[\omega_3\xi_m + (\omega_3\xi_m)^{-1}],$$

$$\gamma_{12}^{(m)}(0) = \sqrt{\beta}\,\gamma_m(0)[\omega_2\xi_m + (\omega_2\xi_m)^{-1}],$$

$$\gamma_{13}^{(m)} = 0,$$

$$\lambda_1(\zeta_1^{(m+1)}) = -\mathrm{i}\sqrt{\beta/3}\,[\omega_3\xi_m + (\omega_3\xi_m)^{-1}],$$

$$\lambda_3(\zeta_1^{(m+1)}) = -\mathrm{i}\sqrt{\beta/3}\,[\omega_2\xi_m + (\omega_2\xi_m)^{-1}],$$

$$\gamma_{12}^{(m+1)} = 0,$$

$$\gamma_{13}^{(m+1)}(0) = \sqrt{\beta}\,\gamma_m(0)[\omega_3\xi_m + (\omega_3\xi_m)^{-1}].$$

For the N-soliton solution there are N arbitrary constants ξ_m and N arbitrary constants γ_m. We note that comparison of $W = 6(\ln f)_X$ (6.2.2) with (7.3.39) shows that

$$\ln(\det \mathbf{M}) = 2\ln f \tag{7.3.41}$$

so that $\det \mathbf{M}$ should be a perfect square for arbitrary N.

Finally, the N-soliton solution of the untransformed GVE (7.3.1) is given in parametric form by

$$u(x,t) = U(t,T), \qquad x = \theta(t,T), \tag{7.3.42}$$

where $\theta(X,T)$ is defined in (7.3.4).

7.3.4 Examples of one- and two-soliton solutions

In order to obtain the one-soliton solution of the GVPE (7.3.2) we need first to calculate the 2×2 matrix $\mathbf{M}$ according to (7.3.40) with $N = 1$. The elements of the matrix are

$$\begin{aligned}
M_{11} &= 1 - \frac{\sqrt{\beta}\,\gamma_1}{2k}\left[\omega_2\xi_1 + (\omega_2\xi_1)^{-1}\right]\exp\left[2k(X - cT)\right], \\
M_{12} &= \frac{\sqrt{3}\,\gamma_1 \mathrm{i}}{2}\left\{\frac{[\omega_2\xi_1 + (\omega_2\xi_1)^{-1}]}{[\omega_3\xi_1 + (\omega_3\xi_1)^{-1}]}\right\} \times \\
&\times \exp\left\{2\mathrm{i}\sqrt{\beta/3}\,[\omega_3\xi_1 + (\omega_3\xi_1)^{-1}]X - 2kcT\right\}, \\
M_{21} &= -\frac{\sqrt{3}\,\gamma_1 \mathrm{i}}{2}\left\{\frac{[\omega_3\xi_1 + (\omega_3\xi_1)^{-1}]}{[\omega_2\xi_1 + (\omega_2\xi_1)^{-1}]}\right\} \times \\
&\times \exp\left\{-2\mathrm{i}\sqrt{\beta/3}\,[\omega_2\xi_1 + (\omega_2\xi_1)^{-1}]X - 2kcT\right\}, \\
M_{22} &= 1 - \frac{\sqrt{\beta}\,\gamma_1}{2k}\left[\omega_3\xi_1 + (\omega_3\xi_1)^{-1}\right]\exp\left[2k(X - cT)\right],
\end{aligned} \tag{7.3.43}$$

and the determinant of the matrix is

$$\det M = \left\{1 + \frac{\gamma_1}{2}\left(\frac{\xi_1^2 + 1}{\xi_1^2 - 1}\right)\exp\left[2k(X - cT)\right]\right\}^2, \tag{7.3.44}$$

where $k = \sqrt{\beta}\,(\xi_1 - \xi_1^{-1})/2$ and $c^{-1} = \beta(\xi_1^2 + \xi_1^{-2} - 1)$. Notice that this determinant is a perfect square; this is consistent with (7.3.41).

From (7.3.39) and (7.3.44), the one-soliton solution of the GVPE (7.3.2), as obtained by the IST method, is

$$U(X, T) = 6k^2\,\mathrm{sech}^2\left[k(X - cT) + \alpha_1\right], \tag{7.3.45}$$

where

$$\alpha_1 = \frac{1}{2}\ln\left[\frac{\gamma_1}{2}\left(\frac{\xi_1^2 + 1}{\xi_1^2 - 1}\right)\right].$$

α_1 is an arbitrary constant. Since U is real, it follows from (7.3.45) that α_1 is real; moreover, since $\sqrt{\beta}\,\xi_1$ is real, γ_1 is also real. (7.3.45) agrees with the one-soliton solution to the GVPE as found by Hirota's method in Section 6.6 and given by (4.1) – (4.4) in [101].

In a similar way (details omitted) we find that, for the two-soliton solution, $\mathbf{M}$ is a 4×4 matrix for which $\det \mathbf{M}$ is a perfect square given by

$$\det \mathbf{M} = \left(1 + q_1^2 + q_2^2 + b_{12}^2 q_1^2 q_2^2\right)^2, \tag{7.3.46}$$

where

$$q_i = \exp\left[2k_i(X - c_i T) + \alpha_i\right], \tag{7.3.47}$$

$$b_{12}^2 = \left(\frac{\xi_3 - \xi_3^{-1} - \xi_1 + \xi_1^{-1}}{\xi_3 - \xi_3^{-1} + \xi_1 - \xi_1^{-1}}\right)^3 \frac{\xi_3^3 - \xi_3^{-3} + \xi_1^3 - \xi_1^{-3}}{\xi_3^3 - \xi_3^{-3} - \xi_1^3 + \xi_1^{-3}}, \tag{7.3.48}$$

$$k_i = \sqrt{\beta}\,(\xi_{2i-1} - \xi_{2i-1}^{-1})/2, \quad c_i^{-1} = \beta(\xi_{2i-1}^2 + \xi_{2i-1}^{-2} - 1),$$

$$\alpha_i = \frac{1}{2}\ln\left[\frac{\gamma_{2i-1}}{2}\left(\frac{\xi_{2i-1}^2 + 1}{\xi_{2i-1}^2 - 1}\right)\right].$$

The α_i are real arbitrary constants. The relationship (7.3.39) together with (7.3.46) gives the two-soliton solution of (7.3.2). (7.3.46), (7.3.47) and (7.3.48) agree with the two-soliton solution as found by Hirota's method in Section 6.6 and given by (7.1) – (7.7) in [101].

In passing we note that in the limit $\beta \to 0$ with $\sqrt{\beta}\,\xi_i$ held constant, the one- and two-soliton solutions given above reduce to the ones obtained in [84] for the VPE.

Consequently, in Section 6.6 (see [101, 102] too) the Vaknenko-Parkes equation has been generalized to an equation that is known as the generalised Vakhnenko-Parkes equation. It turns out that this new evolution equation possesses a wider variety of solutions, is integrable, and has been solved by both the Hirota method in Section 6.6 (see [101, 102] too) and the IST method in Section 7.3 (see [89, 90] too). Now this equation is investigated very actively in the scientific literature.

Chapter 8
Accounting for the continuum part of spectral data

Now, in addition to the bound state spectrum, we consider the continuous spectrum of the associated eigenvalue problem [85, 86, 87, 88], i.e. we assume that at least some of the functions $Q_{1j}(\zeta')$ are nonzero. At each fixed $j \neq 1$, the functions $Q_{1j}(\zeta')$ characterize the singularity of $\Phi_1(X, \zeta)$. As we have shown, this singularity can appear only on boundaries between the regular regions on the ζ-plane, where the condition $\mathrm{Re}(\lambda_1(\zeta') - \lambda_j(\zeta')) = 0$ constitutes these boundaries [132]. For the VPE (5.1.10)

$$W_{XXT} + (1 + W_T)W_X = 0,$$

as we know, the complex ζ-plane is divided into four regions by two lines (7.1.25)

$$\text{(i)} \quad \zeta' = \omega_2\xi, \quad \text{with} \quad Q_{12}^{(1)}(\zeta') \neq 0, \quad Q_{13}^{(1)}(\zeta') \equiv 0,$$

$$\text{(ii)} \quad \zeta' = -\omega_3\xi, \quad \text{with} \quad Q_{12}^{(2)}(\zeta') \equiv 0, \quad Q_{13}^{(2)}(\zeta') \neq 0,$$

where ξ is real (see Fig. 7.1) and sweeps from $-\infty$ to $+\infty$.

8.1 Special continuous spectrum in δ-function form

Let us consider the singularity functions $Q_{1j}(\zeta')$ on the boundaries, on which the Jost function $\phi_1(X, \zeta)$ is singular, in the form ($m =$

1, 2, ..., M)

$$\left.\begin{aligned} Q^{(1)}_{12}(\zeta') &= -2\pi \mathrm{i} \sum_{m=1}^{M} q^{(2m-1)}_{12} \delta(\zeta' - \zeta'_{2n-1}) \\ Q^{(1)}_{13}(\zeta') &= -2\pi \mathrm{i} \sum_{m=1}^{M} q^{(2m-1)}_{13} \delta(\zeta' - \zeta'_{2n-1}) \equiv 0 \end{aligned}\right\} \text{ on the line } \quad \zeta' = \omega_2 \xi,$$

$$\left.\begin{aligned} Q^{(2)}_{12}(\zeta') &= -2\pi \mathrm{i} \sum_{m=1}^{M} q^{(2m)}_{12} \delta(\zeta' - \zeta'_{2n}) \equiv 0 \\ Q^{(2)}_{13}(\zeta') &= -2\pi \mathrm{i} \sum_{m=1}^{M} q^{(2m)}_{13} \delta(\zeta' - \zeta'_{2n}) \end{aligned}\right\} \text{ on the line } \quad \zeta' = -\omega_3 \xi. \tag{8.1.1}$$

For the singularity functions (8.1.1) and for N pairs of poles, the relationship (7.1.29) is reduced to the form (provisionally the time-dependence is not written)

$$\begin{aligned} \Phi_1(X,\zeta) = 1 &- \sum_{k=1}^{2N} \sum_{j=2}^{3} \gamma^{(k)}_{1j} \frac{\exp\{[\lambda_j(\zeta^{(k)}_1) - \lambda_1(\zeta^{(k)}_1)]X\}}{\lambda_1(\zeta^{(k)}_1) - \lambda_1(\zeta)} \Phi_1(X, \omega_j \zeta^{(k)}_1) \\ &- \sum_{l=1}^{2M} \sum_{j=2}^{3} q^{(l)}_{1j} \frac{\exp\{[\lambda_j(\zeta'_l) - \lambda_1(\zeta'_l)]X\}}{\zeta'_l - \zeta} \Phi_1(X, \omega_j \zeta'_l). \end{aligned} \tag{8.1.2}$$

In Section 7.2 (see [84] too) it is proved that the poles appear in pairs only $\zeta^{(2n-1)}_1 = \mathrm{i}\omega_2 \xi_n$ and $\zeta^{(2n)}_1 = -\mathrm{i}\omega_3 \xi_n$, under the conditions $\gamma^{(2n-1)}_{12} = \omega_2 \beta_n$, $\gamma^{(2n-1)}_{13} = 0$, $\gamma^{(2n)}_{12} = 0$, $\gamma^{(2n)}_{13} = \omega_3 \beta_n$, ($n =$ 1, 2, ..., N). If we consider both the bound state spectrum and the continuous spectrum, the constants β_n are complex values in the general case. The restrictions on the β_n for real solutions $U = W_X$ follow from a separate problem which will be analyzed in Section 8.2.

As follows from the relationships (7.2.4) and (8.1.2), the singularities in the form (8.1.1) appear in pairs $\zeta'_{2m-1} = \omega_2 \xi_m$ and $\zeta'_{2m} = -\omega_3 \xi_m$. From (7.2.4), on considering the limits $\zeta \to \zeta'_l$ and $X \to -\infty$, it immediately follows that

$$q^{(2m-1)}_{12} \omega_2 = q^{(2m)}_{13} \quad \text{for} \quad m = 1,\, 2, \ldots, M. \tag{8.1.3}$$

Insofar as we have $2N$ poles and $2M$ coefficients $q^{(2m-1)}_{12}$ and $q^{(2m)}_{13}$ in the adopted specifications (8.1.1) of the singularity functions

$Q_{1j}(\zeta')$, it is convenient to introduce the notation

$$\mu_{ji} = \begin{cases} \lambda_j(\zeta_1^{(i)}) \\ \lambda_j(\zeta'_{(i-K)}) \end{cases}, \quad p_{1j}^{(i)} = \begin{cases} \gamma_{1j}^{(i)} & \text{at} \quad i = 1, ..., K \\ q_{1j}^{(i-K)} & \text{at} \quad i = K+1, ..., K+L \end{cases}, \tag{8.1.4}$$

where $K = 2N$ and $L = 2M$. Then the relationship (8.1.2) is rewritten as follows:

$$\Phi_1(X,\zeta) = 1 - \sum_{i=1}^{K+L} \sum_{j=2}^{3} p_{1j}^{(i)} \frac{\exp[(\mu_{ji} - \mu_{1i})X]}{\mu_{1i} - \zeta} \Phi_1(X, \mu_{ji}). \tag{8.1.5}$$

By defining

$$\Psi_i(X) = \sum_{j=2}^{3} p_{1j}^{(i)} \exp(\mu_{ji} X) \Phi_1(X, \mu_{ji}), \tag{8.1.6}$$

we may rewrite the relationship (8.1.5) as

$$\Phi_1(X,\zeta) = 1 - \sum_{i=1}^{K+L} \frac{\exp(-\mu_{1i} X)}{\mu_{1i} - \zeta} \Psi_i(X). \tag{8.1.7}$$

Taking into account (7.1.30), namely

$$\Phi_1(X,\zeta) = 1 - \frac{1}{3\lambda_1(\zeta)} [W(X) - W(-\infty)] + O(\lambda_1^{-2}(\zeta)),$$

and (8.1.6) and (8.1.7), the following key relationship may be found (see also Eq. (7.2.9))

$$W(X) - W(-\infty) = -3 \sum_{k=1}^{L+M} \exp(-\mu_{jk} X) \Psi_k(X) = 3 \frac{\partial}{\partial X} \ln(\det M(X)). \tag{8.1.8}$$

Here the matrix $\mathbf{M}(X)$ is defined as follows:

$$M_{il}(X) = \delta_{il} - \sum_{j=2}^{3} p_{1j}^{(i)} \frac{\exp[(\mu_{ji} - \mu_{1l})X]}{\mu_{ji} - \mu_{1l}}. \tag{8.1.9}$$

Restoring the T-evolution in the relationships, the final result for the solution of the VPE (5.1.11), when we consider the spectral data from both the bound state spectrum and the continuous spectrum, is as follows:

$$U(X,T) = W_X(X,T) = 3\frac{\partial^2}{\partial X^2}\ln\left(\det \mathbf{M}(X,T)\right). \tag{8.1.10}$$

Here $\mathbf{M}(X,T)$ is the $(K+L)\times(K+L)$ matrix given by

$$M_{kl}(X,T) = \delta_{kl} - \tag{8.1.11}$$

$$-\sum_{j=2}^{3} p_{1j}^{(k)} \frac{\exp\{(\mu_{jk}-\mu_{1l})X + [-(3\mu_{jk})^{-1} + (3\mu_{1k})^{-1}]T\}}{\mu_{jk}-\mu_{1l}},$$

where, for $i \le N$,

$$\begin{array}{ll} \mu_{1(2i-1)} = \lambda_1(\zeta_1^{(2i-1)}) = \mathrm{i}\omega_2\xi_i, & \mu_{2(2i-1)} = \lambda_2(\zeta_1^{(2i-1)}) = \mathrm{i}\omega_3\xi_i, \\ p_{12}^{(2i-1)} = \gamma_{12}^{(2i-1)} = \omega_2\beta_i, & p_{13}^{(2i-1)} = \gamma_{13}^{(2i-1)} = 0, \\ \mu_{1(2i)} = \lambda_1(\zeta_1^{(2i)}) = -\mathrm{i}\omega_3\xi_i, & \mu_{3(2i)} = \lambda_3(\zeta_1^{(2i)}) = -\mathrm{i}\omega_2\xi_i, \\ p_{12}^{(2i)} = \gamma_{12}^{(2i)} = 0, & p_{13}^{(2i)} = \gamma_{13}^{(2i)} = \omega_3\beta_i, \end{array} \tag{8.1.12}$$

and for $N < i \le N+M$,

$$\begin{array}{ll} \mu_{1(2i-1)} = \lambda_1(\zeta'_{2(i-M)-1}) = \omega_2\xi_i, & \mu_{2(2i-1)} = \lambda_2(\zeta'_{2(i-M)-1}) = \omega_3\xi_i, \\ p_{12}^{(2i-1)} = q_{12}^{(2(i-M)-1)} = \omega_2\beta_i, & p_{13}^{(2i-1)} = q_{13}^{(2(i-M)-1)} = 0, \\ \mu_{1(2i)} = \lambda_1(\zeta'_{2(i-M)}) = -\omega_3\xi_i, & \mu_{3(2i)} = \lambda_3(\zeta'_{2(i-M)}) = -\omega_2\xi_i, \\ p_{12}^{(2i)} = q_{12}^{(2(i-M))} = 0, & p_{13}^{(2i)} = q_{13}^{(2(i-M))} = \omega_3\beta_i. \end{array} \tag{8.1.13}$$

For the solution (8.1.10), (8.1.11) there are $(N+M)$ arbitrary constants ξ_i and $(N+M)$ arbitrary constants β_i. The constants ξ_i are real, while the constants β_i, in the general case, are complex.

As will be clear from the examples in Section 8.2, the solution (8.1.10), (8.1.11) includes M discrete frequencies from the continuum part of the spectral data. For this reason, the solution (8.1.10), (8.1.11) without solitons (i.e. with $N = 0$) will be referred to as an M-mode solution of the VPE (5.1.11). Evidently these discrete modes emanate from the special choice (8.1.1) of the singularity functions $Q_{1j}(\zeta')$.

The solution obtained through the matrix (8.1.11) is in general a complex function. Consequently, there is a problem in selecting the real solutions from the complex solutions. It turns out that we can obtain the real solutions by means of restriction of arbitrariness in the choice of the constants β_i. We have succeeded in finding these restrictions.

8.2 Real solutions for the VPE

Now we select the real solutions $U = W_X$ from (8.1.10), (8.1.11). We analyze a number of examples, as well as the general case, for the interaction of the solitons and multi-mode waves [85, 86, 87, 88]. To obtain the solutions of the VPE, one has to calculate the determinant of the matrix (8.1.11). Firstly, we present four results of such a calculation for $N + M \leq 4$. For convenience we will use the auxiliary function $F(X,T)$ given by the definition $F(X,T) = \sqrt{\det \mathbf{M}(X,T)}$. In particular, from (8.1.11),

1) for $N + M = 1$ we have

$$F = 1 + c_1 q_1; \tag{8.2.1}$$

2) for $N + M = 2$ we have

$$F = 1 + c_1 q_1 + c_2 q_2 + b_{12} c_1 c_2 q_1 q_2; \tag{8.2.2}$$

3) for $N + M = 3$ we have

$$\begin{aligned} F = \; & 1 + c_1 q_1 + c_2 q_2 + c_3 q_3 + b_{12} c_1 c_2 q_1 q_2 + b_{13} c_1 c_3 q_1 q_3 \\ & + b_{23} c_2 c_3 q_2 q_3 + b_{12} b_{13} b_{23} c_1 c_2 c_3 q_1 q_2 q_3; \end{aligned} \tag{8.2.3}$$

4) for $N + M = 4$ we have

$$\begin{aligned} F = \; & 1 + c_1 q_1 + c_2 q_2 + c_3 q_3 + c_4 q_4 + b_{12} c_1 c_2 q_1 q_2 + b_{13} c_1 c_3 q_1 q_3 \\ & + b_{14} c_1 c_4 q_1 q_4 + b_{23} c_2 c_3 q_2 q_3 + b_{24} c_2 c_4 q_2 q_4 + b_{34} c_3 c_4 q_3 q_4 \\ & + b_{12} b_{13} b_{23} c_1 c_2 c_3 q_1 q_2 q_3 + b_{12} b_{14} b_{24} c_1 c_2 c_4 q_1 q_2 q_4 \\ & + b_{13} b_{14} b_{34} c_1 c_3 c_4 q_1 q_3 q_4 + b_{23} b_{24} b_{34} c_2 c_3 c_4 q_2 q_3 q_4 \\ & + b_{12} b_{13} b_{14} b_{23} b_{24} b_{34} c_1 c_2 c_3 c_4 q_1 q_2 q_3 q_4. \end{aligned} \tag{8.2.4}$$

For $N+M>4$, the explicit expression for the function $F(X,T)$ can be obtained in a similar manner. It is helpful to present the quantities c_i, q_i and b_{ij} involved in the above formulas (8.2.1) – (8.2.4) separately for three distinct cases:

1. The purely solitonic case $(i,j)\le N$ has

$$q_i=\exp(2\theta_i),\quad 2\theta_i=\sqrt{3}\xi_i X-(\sqrt{3}\xi_i)^{-1}T,\quad c_i=\frac{\beta_i}{2\sqrt{3}\xi_i},$$
$$b_{ij}=\left(\frac{\xi_i-\xi_j}{\xi_i+\xi_j}\right)^2\frac{\xi_i^2+\xi_j^2-\xi_i\xi_j}{\xi_i^2+\xi_j^2+\xi_i\xi_j},\quad b_{ij}\ge 0; \tag{8.2.5}$$

2. The case of purely multi-mode waves $N<(i,j)\le N+M$ has

$$q_i=\exp(2\theta_i),\quad 2\theta_i=-\mathrm{i}\sqrt{3}\xi_i X+(\mathrm{i}\sqrt{3}\xi_i)^{-1}T,\quad c_i=\frac{\mathrm{i}\beta_i}{2\sqrt{3}\xi_i},$$
$$b_{ij}=\left(\frac{\xi_i-\xi_j}{\xi_i+\xi_j}\right)^2\frac{\xi_i^2+\xi_j^2-\xi_i\xi_j}{\xi_i^2+\xi_j^2+\xi_i\xi_j},\quad b_{ij}\ge 0; \tag{8.2.6}$$

3. The case of a combination of solitons $(i,i')\le N$ and multi-mode waves $N<(j,j')\le N+M$ has

$$q_i=\exp(2\theta_i),\quad 2\theta_i=\sqrt{3}\xi_i X-(\sqrt{3}\xi_i)^{-1}T,\quad c_i=\frac{\beta_i}{2\sqrt{3}\xi_i},$$
$$q_j=\exp(2\theta_j),\quad 2\theta_j=-\mathrm{i}\sqrt{3}\xi_j X+(\mathrm{i}\sqrt{3}\xi_j)^{-1}T,\quad c_j=\frac{\mathrm{i}\beta_j}{2\sqrt{3}\xi_j},$$
$$b_{ii'}=\left(\frac{\xi_i-\xi_{i'}}{\xi_i+\xi_{i'}}\right)^2\frac{\xi_i^2+\xi_{i'}^2-\xi_i\xi_{i'}}{\xi_i^2+\xi_{i'}^2+\xi_i\xi_{i'}},\quad 0\le b_{ii'}\le 1,$$
$$b_{jj'}=\left(\frac{\xi_j-\xi_{j'}}{\xi_j+\xi_{j'}}\right)^2\frac{\xi_j^2+\xi_{j'}^2-\xi_j\xi_{j'}}{\xi_j^2+\xi_{j'}^2+\xi_j\xi_{j'}},\quad 0\le b_{jj'}\le 1,$$
$$b_{ij}=\left(\frac{\xi_i+\mathrm{i}\xi_j}{\xi_i-\mathrm{i}\xi_j}\right)^2\frac{\xi_i^2-\xi_j^2+\mathrm{i}\xi_i\xi_j}{\xi_i^2-\xi_j^2-\mathrm{i}\xi_i\xi_j},\quad |b_{ij}|\equiv 1. \tag{8.2.7}$$

With the above found representation of the auxiliary function $F(X,T)$ and taking into account the key relationship (8.1.10), we can write the explicit solution to the basic nonlinear evolution equation (5.1.10) in the following concise form:

$$W(X,T) = 6\frac{\partial}{\partial X}\ln(F(X,T)) + \text{const.} \tag{8.2.8}$$

The function F is complex-valued in the general case because the values of β_i (and hence of c_i) are complex constants.

Since we are interested only in the real solution W_X with real constants ξ_i, we need restrictions on the constants c_i in (8.2.1) – (8.2.4).

8.3 The solutions associated with the continuous spectrum

We study the multi-mode solutions for $N = 0$ and $M = 1, 2, 3, 4$, while for $M \geq 5$ all formulas can easily be obtained by means of a generalization of these examples.

8.3.1 The one-mode solution

In order to obtain the one-mode solution of the VPE (5.1.10) we need first to calculate the 2×2 matrix $\mathbf{M}(X,T)$ according to (8.1.11) with $N = 0$ and $M = 1$. For the matrix elements $M_{kl}(X,T)$ we have

$$\begin{aligned}
M_{11}(X,T) &= 1 - \frac{\mathrm{i}\omega_2\beta_1}{\sqrt{3}\xi_1}\exp[-\mathrm{i}\sqrt{3}\xi_1 X + (\mathrm{i}\sqrt{3}\xi_1)^{-1}T],\\
M_{12}(X,T) &= -\frac{\omega_3\beta_1}{2\xi_1}\exp[2\omega_3\xi_1 X + (\mathrm{i}\sqrt{3}\xi_1)^{-1}T],\\
M_{21}(X,T) &= \frac{\omega_2\beta_1}{2\xi_1}\exp[-2\omega_2\xi_1 X + (\mathrm{i}\sqrt{3}\xi_1)^{-1}T],\\
M_{22}(X,T) &= 1 - \frac{\mathrm{i}\omega_3\beta_1}{\sqrt{3}\xi_1}\exp[-\mathrm{i}\sqrt{3}\xi_1 X + (\mathrm{i}\sqrt{3}\xi_1)^{-1}T],
\end{aligned} \tag{8.3.1}$$

so that the respective determinant is

$$\det \mathbf{M}(X,T) = \left[1 + c_1\exp(-\mathrm{i}\sqrt{3}\xi_1 X + (\mathrm{i}\sqrt{3}\xi_1)^{-1}T)\right]^2, \quad c_1 = \frac{\mathrm{i}\beta_1}{2\sqrt{3}\xi_1}. \tag{8.3.2}$$

As has been noted already, the singularity functions in the form (8.1.1) with $M = 1$ give rise to a single frequency for the continuous part of the spectral data. Hence, the expression (8.3.2), having been substituted into the concise formula (8.2.8), must provide us with the one-mode solution.

The condition that W_X is real requires a restriction on the constant β_1 (if the constant ξ_1 is arbitrary but real). We have succeeded in obtaining this restriction (see Appendix A at end of this Chapter), namely that the constant c_1, which in general is a complex valued one with $c_1 = |c_1|\exp(\mathrm{i}\chi_1)$, should possess unit modulus $|c_1| = 1$, while the arbitrary real constant χ_1 defines an initial shift of solution $X_1 = \chi_1/(\sqrt{3}\xi_1)$ so that

$$\det \mathbf{M}(X,T) = \left[1 + \exp\left(-\mathrm{i}\sqrt{3}\xi_1(X - X_1) + \frac{T}{\mathrm{i}\sqrt{3}\xi_1}\right)\right]^2. \quad (8.3.3)$$

The final result for one mode of the continuous spectrum is the solution (8.2.8) with (8.3.3), namely

$$W(X,T) = -3\sqrt{3}\,\xi_1 \tan\left(\frac{\sqrt{3}}{2}\xi_1(X - X_1) + \frac{T}{2\sqrt{3}\xi_1}\right) + \text{const}. \quad (8.3.4)$$

The corresponding solution for $U = W_X$ was obtained recently by other methods, for example, by the sine-cosine method [142], the (G'/G)-expansion method [78], and the extended tanh-function method [142, 143]. However, only the approach developed here and the solution in the form (8.1.10), (8.1.11) enable us to study the interaction of solitons and periodic waves.

We obtain periodic solutions even for $M = 1$. Let us call attention once again to the condition (7.1.17) which in the final result is shown below to restrict the region of X for periodic solutions. At first glance it would seem that there is a contradiction between the condition (7.1.17) and the periodic solution. Indeed, on the one hand, the condition (7.1.17) demands that the solution $W(X,T)$ should vanish as $X \to -\infty$; on the other hand, the periodic solution obtained here does not satisfy the condition (7.1.17). Nevertheless, consideration of the details enables us to find a reasonable explanation. So, in [132], for the derivation of the relation (7.1.29) (see also (4.5) in [132]), the integral in (7.1.29) appears as a result of the integration on two sides of the boundaries between regular regions. For an understanding of this fact, the relationship (8.1.10) from [132] plays an important role.

Hence, the integration in (7.1.29) (also as in (4.5) in [132]) should be carried out over the lines $\omega_2(\xi + \mathrm{i}\varepsilon)$ and $-\omega_3(\xi + \mathrm{i}\varepsilon)$ as ξ sweeps from $-\infty$ to ∞, where $\varepsilon > 0$. As a result, in the relationship (8.3.3) we should exchange ξ_1 for $(\xi_1 + \mathrm{i}\varepsilon)$ and that enables us to define the solution in the form

$$W(X,T) = \tag{8.3.5}$$

$$= -\mathrm{i}6\sqrt{3}\,(\xi_1 + \mathrm{i}\varepsilon)\frac{\exp(\sqrt{3}\varepsilon X)\exp\left(-\mathrm{i}\sqrt{3}\xi_1(X - X_1) + \frac{T}{\mathrm{i}\sqrt{3}\xi_1}\right)}{1 + \exp(\sqrt{3}\varepsilon X)\exp\left(-\mathrm{i}\sqrt{3}\xi_1(X - X_1) + \frac{T}{\mathrm{i}\sqrt{3}\xi_1}\right)},$$

which tends to constants as $|X| \to \infty$ at arbitrary $\varepsilon > 0$. Thus, on the one hand, the condition (7.1.17) is satisfied, and, on the other hand, at small $\varepsilon > 0$ we have a sufficiently large region over X where the solution associated with a finite $\varepsilon > 0$ and the periodic solution associated with $\varepsilon = 0$ are sufficiently close to each other. The region of X with periodic solutions can be extended to sufficiently large, but finite, $|X|$. For any sequence $\varepsilon_n \to 0$, we remain within the inverse scattering theory [132] where the condition (7.1.17) is not violated. Consequently, the periodic solution obtained at $\varepsilon = 0$ is to be interpreted as the solution of the VPE which is valid on arbitrary but finite $|X|$.

8.3.2 The two-mode solution

Let us consider a two-mode solution of the VPE. In this case $\mathbf{M}(X,T)$ is a 4×4 matrix. For its determinant, according to (8.2.2) we find

$$\sqrt{\det \mathbf{M}(X,T)} = F(X,T) = 1 + c_1 q_1 + c_2 q_2 + b_{12} c_1 c_2 q_1 q_2, \tag{8.3.6}$$

where q_i, c_i, and b_{12} are defined by (8.2.6).

Since the solution W_X should be real and the constants ξ_i are arbitrary, but real, there are restrictions on the constants $c_i = |c_i|\exp(\mathrm{i}\chi_i)$. The real constants χ_i define the initial shifts of solutions $X_i = \chi_i/(\sqrt{3}\xi_i)$. The analysis in considerable detail shows (see Appendix A at end of this Chapter) that the relations $|c_1| = |c_2| = 1/\sqrt{b_{12}}$ are the sufficient conditions in order that W_X be real. Thus, the interaction of two periodic waves for the VPE is described by the relationship (8.2.8) with

$$F(X,T) = 1 + \frac{1}{\sqrt{b_{12}}} q_1 + \frac{1}{\sqrt{b_{12}}} q_2 + q_1 q_2, \tag{8.3.7}$$

where b_{12} is as in (8.2.6), and the dependencies in q_i now contain the phaseshifts $X_i = \chi_i/(\sqrt{3}\xi_i)$ as follows:

$$q_i = \exp(-\mathrm{i}\sqrt{3}\xi_i(X - X_i) + (\mathrm{i}\sqrt{3}\xi_i)^{-1}T). \tag{8.3.8}$$

8.3.3 The three-mode solution

For $N = 0$ and $M = 3$, in the relationship

$$\begin{aligned} F(X,T) = \quad & 1 + c_1q_1 + c_2q_2 + c_3q_3 + c_1c_2b_{12}q_1q_2 + c_1c_3b_{13}q_1q_3 \\ & + c_2c_3b_{23}q_2q_3 + c_1c_2c_3b_{12}b_{13}b_{23}q_1q_2q_3 \end{aligned} \tag{8.3.9}$$

obtained from (8.1.11) (see, also (8.2.3)) with q_i, c_i, and b_{ij} as in (8.2.6), we write $c_i = |c_i|\exp(\mathrm{i}\chi_i)$. Then the arguments χ_i determine the initial phaseshifts of modes $X_i = \chi_i/(\sqrt{3}\xi_i)$. As is proved in Appendix A at end of this Chapter, the conditions on the constants c_i are

$$|c_1| = 1/\sqrt{b_{12}b_{13}}, \qquad |c_2| = 1/\sqrt{b_{12}b_{23}}, \qquad |c_3| = 1/\sqrt{b_{13}b_{23}}. \tag{8.3.10}$$

Hence the three-mode solution is the relation (8.2.8) with

$$\begin{aligned} F(X,T) = \quad & 1 + \frac{1}{\sqrt{b_{12}b_{13}}}(q_1 + q_2q_3) + \frac{1}{\sqrt{b_{12}b_{23}}}(q_2 + q_1q_3) \\ & + \frac{1}{\sqrt{b_{13}b_{23}}}(q_3 + q_1q_2) + q_1q_2q_3. \end{aligned} \tag{8.3.11}$$

Here the phaseshifts X_i are taken into account in q_i by way of (8.3.8).

8.3.4 The four-mode solution

For $N = 0$ and $M = 4$, the restrictions have the form (see Appendix A at end of this Chapter)

$$|c_i| = \prod_{\substack{j=1 \\ j\neq i}}^{4} b_{ij}^{-\frac{1}{2}}, \quad 0 \le b_{ij} = b_{ji} \le 1, \quad i = 1,\, 2,\, 3,\, 4. \tag{8.3.12}$$

The function F for the real solution (8.2.8) is

$$\begin{aligned} F(X,T) = \; & 1+\frac{1}{\sqrt{b_{12}b_{13}b_{14}}}(q_1+q_2q_3q_4)+\frac{1}{\sqrt{b_{12}b_{23}b_{24}}}(q_2+q_1q_3q_4) \\ & +\frac{1}{\sqrt{b_{13}b_{23}b_{34}}}(q_3+q_1q_2q_4)+\frac{1}{\sqrt{b_{14}b_{24}b_{34}}}(q_4+q_1q_2q_3) \\ & +\frac{1}{\sqrt{b_{13}b_{14}b_{23}b_{24}}}(q_1q_2+q_3q_4)+\frac{1}{\sqrt{b_{12}b_{14}b_{23}b_{34}}}(q_1q_3+q_2q_4) \\ & +\frac{1}{\sqrt{b_{12}b_{13}b_{24}b_{34}}}(q_1q_4+q_2q_3)+q_1q_2q_3q_4. \end{aligned} \tag{8.3.13}$$

As before, the b_{ij} and q_i are defined by (8.2.6) and (8.3.8), respectively.

8.4 The solutions associated with bound state spectrum

The features of the solutions associated with the bound state spectrum can be shown by considering the two-soliton solution for which $N = 2$, $M = 0$. The solution (8.2.8) can be obtained through (8.2.2) with (8.2.5), i.e.

$$F(X,T) = 1 + c_1q_1 + c_2q_2 + b_{12}c_1c_2q_1q_2 \tag{8.4.1}$$

with

$$\begin{aligned} & q_i = \exp(2\theta_i), \quad 2\theta_i = \sqrt{3}\xi_i X - (\sqrt{3}\xi_i)^{-1}T, \quad c_i = \frac{\beta_i}{2\sqrt{3}\xi_i}, \\ & b_{ij} = \left(\frac{\xi_i-\xi_j}{\xi_i+\xi_j}\right)^2 \frac{\xi_i^2+\xi_j^2-\xi_i\xi_j}{\xi_i^2+\xi_j^2+\xi_i\xi_j}, \quad b_{ij} \geq 0. \end{aligned} \tag{8.4.2}$$

In Appendix B at end of this Chapter it is proved that the constants c_i have to be real. Moreover, the signs of $\alpha_i = c_i/|c_i|$ can independently take the values ± 1, i.e. we have four variants, namely $\alpha_1 = \alpha_2 = 1$, $\alpha_1 = \alpha_2 = -1$, $\alpha_1 = -\alpha_2 = 1$, and $\alpha_1 = -\alpha_2 = -1$. Note that in [123] only the first two variants are discussed. The standard soliton solution for which $\alpha_1 = \alpha_2 = 1$, and the singular soliton solutions for which $\alpha_1 = \alpha_2 = -1$, $\alpha_1 = -\alpha_2 = 1$ and $\alpha_1 = -\alpha_2 = -1$, are obtained by means of the relation (8.2.8) to give

$$U(X,T) = W(X,T)_X = 6\frac{\partial^2}{\partial X^2}\ln(F) = 6\frac{\partial^2}{\partial X^2}\ln(G_i), \tag{8.4.3}$$

where G_i are defined by (B.6) – (B.9).

The forms (B.3), (B.6) – (B.9) for F are more preferable, since we see that the solution is dependent on two combinations of the spectral parameters $\xi_1+\xi_2$ and $\xi_1-\xi_2$, but not three values ξ_1, ξ_2, and $\xi_1+\xi_2$ as it may appear from the relation (8.4.3).

For $N \geq 3$ we give the conditions without proof. All the constants c_i are to be real and the signs of $\alpha_i = c_i/|c_i|$ can be equal to ± 1 independently of each other.

8.5 Real soliton and multi-mode solutions of the VPE

In this subsection we will consider the general case, when both the bound state spectrum and the continuous spectrum are taken into account in the associated spectral problem. We will find the conditions on c_i for real solutions of the VPE. To obtain the solution, we need to know the function F (see (8.2.1) – (8.2.4)).

Let the indexes i and i' be related to the values involved in the bound state spectrum for which $(i,\, i') \leq N$, while the indexes j and j' are related to the values involved in the continuous part of the spectral data for which $N < (j, j') \leq N+M$.

8.5.1 The interaction of a soliton with a one-mode wave

The interaction of a standard soliton with a periodic one-mode wave can be described by means of the relation (8.2.2) with $N=1$ and $M=1$, namely

$$F(X,T) = 1 + c_1 q_1 + c_2 q_2 + b_{12} c_1 c_2 q_1 q_2 \tag{8.5.1}$$

with q_i and b_{12} as in (8.2.7), namely

$$\begin{aligned} q_1 &= \exp(\sqrt{3}\xi_1 X - (\sqrt{3}\xi_1)^{-1} T), \quad c_1 = \frac{\beta_1}{2\sqrt{3}\xi_1}, \\ q_2 &= \exp(-\mathrm{i}\sqrt{3}\xi_2 X + (\mathrm{i}\sqrt{3}\xi_2)^{-1} T), \quad c_2 = \frac{\mathrm{i}\beta_2}{2\sqrt{3}\xi_2}, \\ b_{12} &= \left(\frac{\xi_1 + \mathrm{i}\xi_2}{\xi_1 - \mathrm{i}\xi_2}\right)^2 \frac{\xi_1^2 - \xi_2^2 + \mathrm{i}\xi_1\xi_2}{\xi_1^2 - \xi_2^2 - \mathrm{i}\xi_1\xi_2}, \quad |b_{12}| \equiv 1. \end{aligned} \tag{8.5.2}$$

First, we emphasize that the soliton and one-mode wave (8.3.4) propagate in opposite directions. The soliton propagates in the positive direction of the X-axis, while the one-mode wave (8.3.4) propagates in the negative direction of the X-axis.

Here we restrict ourselves to the simplest case $b_{12}c_1c_2 = 1$ that describes the interaction of a standard soliton with a one-mode wave. As follows immediately from Appendix C at end of this Chapter, for real solutions (8.2.8),

$$W(X,T) = 6\frac{\partial}{\partial X}\ln(F(X,T)) + \text{const.},$$

where $F(X,T)$ is

$$F(X,T) = 1 + \frac{1}{\sqrt{b_{12}}}q_1 + \frac{1}{\sqrt{b_{12}}}q_2 + q_1q_2. \tag{8.5.3}$$

There is an exceptional case at $\xi_1 = \xi_2$. Then we have $b_{12} = 1$, and $F = (1+q_1)(1+q_2)$. Consequently, the solution (8.2.8) is reduced to the relation

$$\begin{aligned} W = W_1 + W_2 = \; & 3\sqrt{3}\xi_1 \tanh\left(\frac{\sqrt{3}}{2}\xi_1(X - X_1) - \frac{T}{2\sqrt{3}\xi_1}\right) \\ & -3\sqrt{3}\,\xi_1 \tan\left(\frac{\sqrt{3}}{2}\xi_1(X - X_0) + \frac{T}{2\sqrt{3}\xi_1}\right) + \text{const.} \end{aligned} \tag{8.5.4}$$

Here W_1 is the one-soliton solution and W_2 is the solution (8.3.4) associated with one mode in the continuous part of the spectral data. The relationship $W = W_1 + W_2$ is easily verified also by direct substitution into the VPE (5.1.10). The two waves W_1 and W_2 propagate in different directions with the same speed without change of wave profile and phaseshift. In other words, only in the case $\xi_1 = \xi_2$ is there a simple superposition of the solutions W_1 and W_2. It is obvious that interactions of two solitons with a one-mode wave and/or of the two-mode solution with one soliton do not satisfy this form of the interaction.

8.5.2 Real solutions for N solitons and the M-mode wave

The interaction of N solitons and the M-mode wave (8.3.4) can be obtained by means of the function $F(X,T)$ with restrictions (C.7) given in Appendix C, namely

$$c_i = \pm 1 \Big/ \sqrt{\prod_{\substack{j=1 \\ j\neq i}}^{N+M} b_{ij}}\,, \quad b_{ij} = b_{ji}, \quad i = 1, ..., N+M, \qquad (8.5.5)$$

and with the retention of the phaseshifts X_i in the quantities q_i (C.2). The signs for c_i in (8.5.5) can be chosen independently of each other. If the index i in (8.5.5) is connected with the continuous part of the spectral data ($N < i \leq N + M$), then the solutions generated by plus and minus signs in (8.5.5) are different only in the phaseshifts. However, for the index i from the bound state spectrum ($i \leq N$), the solutions have different forms of function dependence. Here it is relevant to remember that there are standard soliton solutions and singular soliton solutions generated by different signs in the constants c_i (8.5.5).

The solution will contain $(N+M)$ real constants ξ_i for determining the values b_{ij} and $(N+M)$ real constants X_i to define the phaseshifts.

We have described the procedure for finding the solutions of the Vakhnenko-Parkes equation by means of the inverse scattering method. Both the bound state spectrum and the continuous spectrum are taken into account in the associated eigenvalue problem. The special form of the singularity functions enables us to obtain the multi-mode solutions. Sufficient conditions have been proved in order that the solutions become real functions. Finally we studied the interaction of solitons and the multi-mode wave.

8.6 Appendices

A. The conditions on the constants c_i for multi-mode waves

In this appendix we will prove the conditions on the constants $c_i = |c_i| \exp(\mathrm{i}\chi_i)$ for solutions associated with the continuous part of the spectral data only. We use the case $M = 4$ as an example to prove the restrictions on the constants, at which the solution $W_X(X,T)$ is real. The auxiliary function $F(X,T)=\sqrt{\det \mathbf{M}(X,T)}$ for finding the solution is (8.2.4), namely

$$\begin{aligned} F(X,T) = \; & 1 + c_1q_1 + c_2q_2 + c_3q_3 + c_4q_4 + c_1c_2b_{12}q_1q_2 + c_1c_3b_{13}q_1q_3 \\ & + c_1c_4b_{14}q_1q_4 + c_2c_3b_{23}q_2q_3 + c_2c_4b_{24}q_2q_4 + c_3c_4b_{34}q_3q_4 \\ & + c_1c_2c_3b_{12}b_{13}b_{23}q_1q_2q_3 + c_1c_2c_4b_{12}b_{14}b_{24}q_1q_2q_4 \\ & + c_1c_3c_4b_{13}b_{14}b_{34}q_1q_3q_4 + c_2c_3c_4b_{23}b_{24}b_{34}q_2q_3q_4 \\ & + c_1c_2c_3c_4b_{12}b_{13}b_{14}b_{23}b_{24}b_{34}q_1q_2q_3q_4. \end{aligned} \tag{A.1}$$

Here we redefine the values c_i in such a way that $c_i = |c_i|$, since the arguments χ_i can always be introduced into the variables $q_i = \exp(\mathrm{i}2\theta_i)$ with $2\theta_i = -\sqrt{3}\xi_i(X - X_i) - (\sqrt{3}\xi_i)^{-1}T$ and $X_i = \chi_i/(\sqrt{3}\xi_i)$ serving as the shifts of solutions. The solution to the VPE (5.1.10) then has the form (8.2.8)

$$W(X,T) = 6\frac{\partial}{\partial X}\ln(F(X,T)) + \text{const.} \tag{A.2}$$

The function F is complex-valued, i.e.

$$F = F_{Re} + \mathrm{i}F_{Im} = |F|\exp(\mathrm{i}\chi_F), \quad F_{Re} = \mathrm{Re}(F), \quad F_{Im} = \mathrm{Im}(F), \tag{A.3}$$

$$\tan(\chi_F) = F_{Im}/F_{Re},$$

hence

$$W(X,T)/6 = \frac{\partial}{\partial X}\ln(|F|) + \mathrm{i}\frac{\partial \chi_F}{\partial X} + \text{const.} \tag{A.4}$$

If we succeed in making $\partial^2\chi_F/\partial X^2 \equiv 0$ by the choice of the constants c_i, then $W_X(X,T)$ will be a real function.

Let us write F_{Im} and F_{Re} in explicit forms, namely

$$\begin{aligned}F_{Im} = \; & c_1 \sin(2\theta_1) + c_2 \sin(2\theta_2) + c_3 \sin(2\theta_3) + c_4 \sin(2\theta_4) \\ & + c_1 c_2 b_{12} \sin[2(\theta_1 + \theta_2)] + c_1 c_3 b_{13} \sin[2(\theta_1 + \theta_3)] \\ & + c_1 c_4 b_{14} \sin[2(\theta_1 + \theta_4)] + c_2 c_3 b_{23} \sin[2(\theta_2 + \theta_3)] \\ & + c_2 c_4 b_{24} \sin[2(\theta_2 + \theta_4)] + c_3 c_4 b_{34} \sin[2(\theta_3 + \theta_4)] \\ & + c_1 c_2 c_3 b_{12} b_{13} b_{23} \sin[2(\theta_1 + \theta_2 + \theta_3)] \\ & + c_1 c_2 c_4 b_{12} b_{14} b_{24} \sin[2(\theta_1 + \theta_2 + \theta_4)] \\ & + c_1 c_3 c_4 b_{13} b_{14} b_{34} \sin[2(\theta_1 + \theta_3 + \theta_4)] \\ & + c_2 c_3 c_4 b_{23} b_{24} b_{34} \sin[2(\theta_2 + \theta_3 + \theta_4)] \\ & + c_1 c_2 c_3 c_4 b_{12} b_{13} b_{14} b_{23} b_{24} b_{34} \sin[2(\theta_1 + \theta_2 + \theta_3 + \theta_4)],\end{aligned} \tag{A.5}$$

$$\begin{aligned}F_{Re} = \; & 1 + c_1 \cos(2\theta_1) + c_2 \cos(2\theta_2) + c_3 \cos(2\theta_3) + c_4 \cos(2\theta_4) \\ & + c_1 c_2 b_{12} \cos[2(\theta_1 + \theta_2)] + c_1 c_3 b_{13} \cos[2(\theta_1 + \theta_3)] \\ & + c_1 c_4 b_{14} \cos[2(\theta_1 + \theta_4)] + c_2 c_3 b_{23} \cos[2(\theta_2 + \theta_3)] \\ & + c_2 c_4 b_{24} \cos[2(\theta_2 + \theta_4)] + c_3 c_4 b_{34} \cos[2(\theta_3 + \theta_4)] \\ & + c_1 c_2 c_3 b_{12} b_{13} b_{23} \cos[2(\theta_1 + \theta_2 + \theta_3)] \\ & + c_1 c_2 c_4 b_{12} b_{14} b_{24} \cos[2(\theta_1 + \theta_2 + \theta_4)] \\ & + c_1 c_3 c_4 b_{13} b_{14} b_{34} \cos[2(\theta_1 + \theta_3 + \theta_4)] \\ & + c_2 c_3 c_4 b_{23} b_{24} b_{34} \cos[2(\theta_2 + \theta_3 + \theta_4)] \\ & + c_1 c_2 c_3 c_4 b_{12} b_{13} b_{14} b_{23} b_{24} b_{34} \cos[2(\theta_1 + \theta_2 + \theta_3 + \theta_4)].\end{aligned} \tag{A.6}$$

Let us try to present F_{Im} and F_{Re} in the forms

$$F_{Im} = 2G \sin(\theta_1 + \theta_2 + \theta_3 + \theta_4) \tag{A.7}$$

and

$$F_{Re} = 2G\cos(\theta_1 + \theta_2 + \theta_3 + \theta_4), \tag{A.8}$$

where G is the same in both the above formulas (A.7) and (A.8). This can be done if the following conditions are satisfied:

$$c_1 = c_2c_3c_4b_{23}b_{24}b_{34}, \quad c_2 = c_1c_3c_4b_{13}b_{14}b_{34}, \quad c_3 = c_1c_2c_4b_{12}b_{14}b_{24},$$
$$c_4 = c_1c_2c_3b_{12}b_{13}b_{23}, \quad c_1c_2b_{12} = c_3c_4b_{34}, \quad c_1c_3b_{13} = c_2c_4b_{24},$$
$$c_1c_4b_{14} = c_2c_3b_{23}, \quad c_1c_2c_3c_4b_{12}b_{13}b_{14}b_{23}b_{24}b_{34} = 1. \tag{A.9}$$

It turns out that all these relations are valid when

$$c_1 = \frac{1}{\sqrt{b_{12}b_{13}b_{14}}}, \quad c_2 = \frac{1}{\sqrt{b_{12}b_{23}b_{24}}}, \quad c_3 = \frac{1}{\sqrt{b_{13}b_{23}b_{34}}}, \quad c_4 = \frac{1}{\sqrt{b_{14}b_{24}b_{34}}}. \tag{A.10}$$

With the conditions (A.10), the expression for G reads as follows:

$$\begin{aligned} G = \ & \cos(\theta_1 + \theta_2 + \theta_3 + \theta_4) + \frac{1}{\sqrt{b_{12}b_{13}b_{14}}}\cos(\theta_1 - \theta_2 - \theta_3 - \theta_4) \\ & + \frac{1}{\sqrt{b_{12}b_{23}b_{24}}}\cos(\theta_2 - \theta_1 - \theta_3 - \theta_4) \\ & + \frac{1}{\sqrt{b_{13}b_{23}b_{34}}}\cos(\theta_3 - \theta_1 - \theta_2 - \theta_4) \\ & + \frac{1}{\sqrt{b_{14}b_{24}b_{34}}}\cos(\theta_4 - \theta_1 - \theta_2 - \theta_3) \\ & + \frac{1}{\sqrt{b_{13}b_{14}b_{23}b_{24}}}\cos(\theta_1 + \theta_2 - \theta_3 - \theta_4) \\ & + \frac{1}{\sqrt{b_{12}b_{14}b_{23}b_{34}}}\cos(\theta_1 + \theta_3 - \theta_2 - \theta_4) \\ & + \frac{1}{\sqrt{b_{12}b_{13}b_{24}b_{34}}}\cos(\theta_1 + \theta_4 - \theta_2 - \theta_3). \end{aligned} \tag{A.11}$$

Now it is readily seen from (A.3) that

$$\chi_F = \theta_1 + \theta_2 + \theta_3 + \theta_4 \tag{A.12}$$

and as consequence we have

$$\frac{\partial^2 \chi_F}{\partial X^2} = \frac{\partial^2 \chi_F}{\partial X \partial T} = 0. \tag{A.13}$$

Hence, as follows from (A.4), the four-mode solution of the VPE (5.1.11) can be reduced to real form with four real constants X_i and four real constants ξ_i (see (8.3.13)).

Here, without proof, we give the following conditions on the constants c_i that ensure the real M-mode solution of the VPE:

$$|c_i| = \prod_{\substack{j=1 \\ j \neq i}}^{M} b_{ij}^{-\frac{1}{2}}, \quad b_{ij} = b_{ji}, \quad i = 1, ..., M, \tag{A.14}$$

where the M constants ξ_i determine the values b_{ij} and the M constants X_i define the phaseshifts for each mode. Note that these relations (A.14) are sufficient conditions, but not necessary ones.

B. The conditions on the constants c_i under the interaction of two solitons

Here we consider the conditions on signs for the constants c_i under the interaction of two solitons ($N = 2$, $M = 0$). We start with the relationship (8.2.2) and (8.2.5)

$$F = 1 + c_1 q_1 + c_2 q_2 + b_{12} c_1 c_2 q_1 q_2. \tag{B.1}$$

Let us present the constants c_i in the form

$$\begin{aligned} c_i &= \alpha_i |c_i| \exp(\mathrm{i}\chi_i) = b_{12}^{-1/2} \exp\left(-\sqrt{3}\xi_i X_i + \mathrm{i}\sigma_i\right), \\ \sigma_i &= \chi_i + \pi(1 - \alpha_i)/2. \end{aligned} \tag{B.2}$$

All the new constants χ_i and $X_i = -\ln(|c_i \sqrt{b_{12}}|)/(\sqrt{3}\xi_i)$ are real. We assume that $-\pi/2 < \chi_i \leq \pi/2$, then the values α_i retain the signs of the constants $\mathrm{Re}(c_i)$, i.e. $\alpha_i = \mathrm{Re}(c_i)/|\mathrm{Re}(c_i)|$. It is convenient for analyzing to rewrite (B.1) (the same as (8.2.2)) in the form

$$F = 2 \exp\left(\theta_1 + \theta_2 + \frac{\mathrm{i}}{2}(\sigma_1 + \sigma_2)\right) G \tag{B.3}$$

with

$$\begin{aligned} G = &\ \cosh\left(\theta_1 + \theta_2 + \tfrac{\mathrm{i}}{2}(\sigma_1 + \sigma_2)\right) \\ &+ b_{12}^{-1/2} \cosh\left(\theta_1 - \theta_2 + \tfrac{\mathrm{i}}{2}(\sigma_1 - \sigma_2)\right), \end{aligned} \tag{B.4}$$

$$2\theta_i = \sqrt{3}\xi_i(X - X_i) - (\sqrt{3}\xi_i)^{-1}T.$$

It is easily seen that only G defines the solution, since $\dfrac{\partial^2}{\partial X^2}\ln(F) = \dfrac{\partial^2}{\partial X^2}\ln(G)$, while the conditions that the function G is real are as follows:

$$\chi_i = 0, \quad \sigma_i + \sigma_2 = 2\pi k_1, \quad \sigma_i - \sigma_2 = 2\pi k_2 \tag{B.5}$$

with $k_i = 0,\ 1$. These restrictions (B.5) lead to the requirements $\alpha_1 = \pm 1$, $\alpha_2 = \pm 1$, independently of each other, and $\chi_i = 0$. Then the function F has the following forms:

1. for $\alpha_1 = \alpha_2 = 1$

$$\begin{aligned} F &= 2\exp(\theta_1 + \theta_2)\, G_1, \\ G_1 &= \cosh(\theta_1 + \theta_2) + b_{12}^{-1/2} \cosh(\theta_1 - \theta_2); \end{aligned} \tag{B.6}$$

2. for $\alpha_1 = \alpha_2 = -1$

$$\begin{aligned} F &= 2\exp(\theta_1 + \theta_2)\, G_2, \\ G_2 &= \cosh(\theta_1 + \theta_2) - b_{12}^{-1/2} \cosh(\theta_1 - \theta_2); \end{aligned} \tag{B.7}$$

3. for $\alpha_1 = -\alpha_2 = 1$

$$\begin{aligned} F &= 2\exp(\theta_1 + \theta_2)\, G_3, \\ G_3 &= -\sinh(\theta_1 + \theta_2) + b_{12}^{-1/2} \sinh(\theta_1 - \theta_2); \end{aligned} \tag{B.8}$$

4. for $\alpha_1 = -\alpha_2 = -1$

$$\begin{aligned} F &= 2\exp(\theta_1 + \theta_2)\, G_4, \\ G_4 &= -\sinh(\theta_1 + \theta_2) - b_{12}^{-1/2} \sinh(\theta_1 - \theta_2). \end{aligned} \tag{B.9}$$

Hence, the standard soliton solution that follows from (B.6) and the singular soliton solutions that follow from (B.7) – (B.9) are the real functions

$$U(X,T) = W_X(X,T) = 6\frac{\partial^2}{\partial X^2}\ln(G_i). \tag{B.10}$$

Now we rewrite the restrictions in a somewhat different form. By retaining the values of the phaseshifts X_i in the quantities q_i, we require

$$c_1 = \pm 1/\sqrt{b_{12}}\,, \quad c_2 = \pm 1/\sqrt{b_{12}}\,, \tag{B.11}$$

where the signs are independent of each other. Note that for this case there are two arbitrary real constants ξ_i, and two arbitrary real constants X_i $(i = 1, 2)$.

The notation in (B.6) – (B.9) shows that the solution is defined by two combinations of the spectral parameters, namely $\xi_1 + \xi_2$ and $\xi_1 - \xi_2$, but not three values ξ_1, ξ_2, $\xi_1 + \xi_2$ as it may appear from (B.1).

The foregoing proof points to a way for finding the restrictions for any N with $M = 0$. Here it should be underlined that only at real c_i with any sign of $\alpha_i = c_i/|c_i|$, the soliton (or singular soliton) solutions are determined by a real function. The conditions on the constants c_i are as follows:

$$c_i = \pm 1 \Bigg/ \sqrt{\prod_{\substack{j=1\\ j\neq i}}^{N} b_{12}} \qquad i = 1, \ldots, N \tag{B.12}$$

with the retention of the phaseshifts X_i in the quantities q_i. The signs for c_i are independent of each other. The solution will contain the N real constants ξ_i for determining the values b_{ij} and the N real constants X_i to define the phaseshifts.

C. The restrictions on the constants c_i in the general case

In this Appendix we will obtain the restrictions on the constants c_i for real solutions, in the general case, taking into account the spectral data from both the bound state spectrum and the continuous spectrum. All features are inherent in the case $N + M = 4$ considered here as an example. To find the solution by means of the inverse scattering method, one needs to know the function (8.2.4)

$$\begin{aligned} F = \ & 1 + c_1 q_1 + c_2 q_2 + c_3 q_3 + c_4 q_4 + b_{12} c_1 c_2 q_1 q_2 + b_{13} c_1 c_3 q_1 q_3 \\ & + b_{14} c_1 c_4 q_1 q_4 + b_{23} c_2 c_3 q_2 q_3 + b_{24} c_2 c_4 q_2 q_4 + b_{34} c_3 c_4 q_3 q_4 \\ & + b_{12} b_{13} b_{23} c_1 c_2 c_3 q_1 q_2 q_3 + b_{12} b_{14} b_{24} c_1 c_2 c_4 q_1 q_2 q_4 \\ & + b_{13} b_{14} b_{34} c_1 c_3 c_4 q_1 q_3 q_4 + b_{23} b_{24} b_{34} c_2 c_3 c_4 q_2 q_3 q_4 \\ & + b_{12} b_{13} b_{14} b_{23} b_{24} b_{34} c_1 c_2 c_3 c_4 q_1 q_2 q_3 q_4. \end{aligned} \tag{C.1}$$

For convenience we rewrite the variables q_i in the somewhat different form

$$\begin{aligned} & q_i = \exp(2\theta_i), \quad q_j = \exp(\mathrm{i}2\theta_j), \quad 2\theta_i = \sqrt{3}\xi_i(X - X_i) - (\sqrt{3}\xi_i)^{-1}T, \\ & 2\theta_j = -\sqrt{3}\xi_j(X - X_j) - (\sqrt{3}\xi_j)^{-1}T. \end{aligned} \tag{C.2}$$

The phaseshifts X_i are arbitrary real constants. The values b_{ij} in (C.1) are as in (8.2.7), namely

$$\begin{aligned} b_{ii'} &= \left(\frac{\xi_i - \xi_{i'}}{\xi_i + \xi_{i'}}\right)^2 \frac{\xi_i^2 + \xi_{i'}^2 - \xi_i \xi_{i'}}{\xi_i^2 + \xi_{i'}^2 + \xi_i \xi_{i'}}, \quad 0 \le b_{ii'} \le 1, \\ b_{jj'} &= \left(\frac{\xi_j - \xi_{j'}}{\xi_j + \xi_{j'}}\right)^2 \frac{\xi_j^2 + \xi_{j'}^2 - \xi_j \xi_{j'}}{\xi_j^2 + \xi_{j'}^2 + \xi_j \xi_{j'}}, \quad 0 \le b_{jj'} \le 1, \\ b_{ij} &= \left(\frac{\xi_i + \mathrm{i}\xi_j}{\xi_i - \mathrm{i}\xi_j}\right)^2 \frac{\xi_i^2 - \xi_j^2 + \mathrm{i}\xi_i \xi_j}{\xi_i^2 - \xi_j^2 - \mathrm{i}\xi_i \xi_j}, \quad |b_{ij}| \equiv 1, \end{aligned} \tag{C.3}$$

where $(i, i') \le N$, and $N < (j, j') \le N + M$. Note that $b_{ii'}$ and $b_{jj'}$ are real values, and $b_{ij}^* = 1/b_{ij}$.

Without loss of generality, we will consider one set of values N and M, for example $N = 1$ and $M = 3$. Now we will show that the restrictions (A.10)

$$\begin{aligned} & c_1 = \pm 1/\sqrt{b_{12} b_{13} b_{14}}, \quad c_2 = \pm 1/\sqrt{b_{12} b_{23} b_{24}}, \\ & c_3 = \pm 1/\sqrt{b_{13} b_{23} b_{34}}, \quad c_4 = \pm 1/\sqrt{b_{14} b_{24} b_{34}} \end{aligned} \tag{C.4}$$

(with b_{ij} determined by (C.3)) are sufficient in order to obtain the real solutions.

For definiteness, we assume that $\sqrt{b_{ij}}$ is a root of the equation $x^2 = b_{ij}$ with $-\pi/2 < \arg\sqrt{b_{ij}} \le \pi/2$. Let us rewrite the relations (C.4) in the form $c_i = \alpha_i / \prod_{\substack{j=1 \\ j\neq i}}^{4} \sqrt{b_{ij}}$, where $\alpha_i = \pm 1$. It is evident that we can always attain $\alpha_2 = \alpha_3 = \alpha_4 = 1$ by choosing the phaseshifts X_2, X_3 and X_4, while we need to consider the two cases $\alpha_1 = \pm 1$. By defining $\sigma = (1-\alpha_1)/2$, we can rewrite the auxiliary function F from (C.1) in the form

$$F(X,T) = 2Ge^{i\pi\sigma}(b_{12}b_{13}b_{14})^{-1/4}\exp(\theta_1 + i\pi\sigma/2 + i\theta_2 + i\theta_3 + i\theta_4),$$

$$\begin{aligned}
Ge^{i\pi\sigma} = \; & \big[(b_{12}b_{13}b_{14})^{1/4}\cos(-i\theta_1 + \pi\sigma/2 + \theta_2 + \theta_3 + \theta_4) \\
& +(b_{12}b_{13}b_{14})^{-1/4}\cos(-i\theta_1 + \pi\sigma/2 - \theta_2 - \theta_3 - \theta_4)\big] \\
& +(b_{23}b_{24})^{-1/2}\big[(b_{13}b_{14}/b_{12})^{1/4}\cos(i\theta_1 - \pi\sigma/2 + \theta_2 - \theta_3 - \theta_4) \\
& + (b_{13}b_{14}/b_{12})^{-1/4}\cos(-i\theta_1 + \pi\sigma/2 + \theta_2 - \theta_3 - \theta_4)\big] \\
& +(b_{23}b_{34})^{-1/2}\big[(b_{12}b_{14}/b_{13})^{1/4}\cos(i\theta_1 - \pi\sigma/2 + \theta_3 - \theta_2 - \theta_4) \\
& + (b_{12}b_{14}/b_{13})^{-1/4}\cos(-i\theta_1 + \pi\sigma/2 + \theta_3 - \theta_2 - \theta_4)\big] \\
& +(b_{24}b_{34})^{-1/2}\big[(b_{12}b_{13}/b_{14})^{1/4}\cos(i\theta_1 - \pi\sigma/2 + \theta_4 - \theta_2 - \theta_3) \\
& + (b_{12}b_{13}/b_{14})^{-1/4}\cos(-i\theta_1 + \pi\sigma/2 + \theta_4 - \theta_2 - \theta_3)\big] .
\end{aligned} \tag{C.5}$$

Since b_{23}, b_{24}, and b_{34} are real, and $b_{1j}^* = 1/b_{1j}$ for $j = 2,3,4$, it is evident that $G^* = G$, i.e. the variable G in the solution is a real-valued function. Hence the solution of the VPE (5.1.11), namely

$$U(X,T) = W_X(X,T) = 6\frac{\partial^2}{\partial X^2}\ln(F) = 6\frac{\partial^2}{\partial X^2}\ln(G), \tag{C.6}$$

is a real quantity.

Using this example, one can prove without difficulty that the procedure considered above can be extended to any N and M with restrictions (see also (A.14), (B.12), (C.4))

$$c_i = \pm 1 \Bigg/ \sqrt{\prod_{\substack{j=1 \\ j\neq i}}^{N+M} b_{ij}}\,, \quad b_{ij} = b_{ji}, \quad i = 1, ..., N+M, \tag{C.7}$$

while the quantities q_i retain the phaseshifts X_i (see (C.2)). The signs in (C.7) can be chosen independently of each other. For the interaction of N solitons and the M-mode wave there are $(N + M)$ real constants ξ_i and $(N + M)$ real constants X_i.

Note that the restrictions (C.7) are sufficient conditions in order that the solution of the VPE is real.

Chapter 9
From simple poles to multiple poles

In this Chapter, the standard procedure for the inverse scattering transform (IST) method is expanded for the case of multiple poles. It is known that the IST method is one of the fundamental methods for solving various nonlinear evolution equations. The method enables one both to provide a proof of the complete integrability of the equation and to solve the initial value problem for this evolution equation (see Chapters 7 and 8). In the IST method, the equation is written as the compatibility condition for two linear equations (the Lax pair). Then the initial condition is mapped into the scattering data. It is important that the spectrum always retains constant values. The time evolution of scattering data is simple and linear. From a knowledge of scattering data evolution, the solution is reconstructed. Hence, for this method the direct spectral problem and the inverse spectral problem are considered. The latter consists of reconstructing the solution of the nonlinear equation from the spectral data. In the general case it is necessary to analyze both the discrete part and the continuum part of the spectral data. It is well-known that the discrete part is associated with soliton solutions, while the continuum part of the spectral data is related to the periodical solutions.

For the spectrum of bound states, we expand the standard procedure from simple poles to the multiple poles. Moreover, for continuum states, a special form of the spectral data is considered where, in the mathematical sense, the problems for the bound states and for continuum states are similar. The spectrum of continuum states is taken as a line spectrum that in first order approximates the step-function. In Section 9.2 onwards we reconstruct the solution from spectral data of the special form. The scope for the suggested spectral data is demonstrated through the analysis of the Vakhnenko-Parkes equation that allows new solutions to be obtained.

9.1 The spectral problem for simple poles

Based on our experience of the study of the Vakhnenko-Parkes equation (VPE), we acquaint the reader with the expanded procedure for taking into account the multiple poles in the discrete part of the spectral data. Using the VPE [84, 91, 85, 86, 87]

$$W_{XXT} + (1 + W_T)W_X = 0 \tag{9.1.1}$$

or, in alternative form with $U \equiv W_X$,

$$UU_{XXT} - U_X U_{XT} + U^2 U_T = 0$$

as an example, we aim to examine both the two-multiple poles and some special forms of the spectral data for which the inverse problem can be solved. In Section 7.1 (see [84] too) it was proved that the Lax pair has the form

$$\psi_{XXX} + W_X\psi_X - \lambda\psi = 0, \tag{9.1.2}$$

$$3\psi_{XT} + (1 + W_T)\psi + \mu\psi_X = 0. \tag{9.1.3}$$

For convenience and for comparison with new results (see Section 9.2), we repeat some results from Chapter 7 for simple poles. The spectral equation (9.1.2) is known to have a matrix form (7.1.14)

$$\frac{\partial}{\partial X}\boldsymbol{\psi} = [\mathbf{A}(\zeta) + \mathbf{B}(X, \zeta)] \cdot \boldsymbol{\psi} \tag{9.1.4}$$

with

$$\boldsymbol{\psi} = \begin{pmatrix} \psi \\ \psi_X \\ \psi_{XX} \end{pmatrix}, \quad \mathbf{A} = \begin{pmatrix} 0 & 1 & 0 \\ 0 & 0 & 1 \\ \lambda & 0 & 0 \end{pmatrix}, \quad \mathbf{B} = \begin{pmatrix} 0 & 0 & 0 \\ 0 & 0 & 0 \\ 0 & -W_X & 0 \end{pmatrix}. \tag{9.1.5}$$

The matrix $\mathbf{A}$ has eigenvalues $\lambda_j(\zeta)$ and left- and right-eigenvectors $\tilde{\boldsymbol{v}}_j(\zeta)$ and $\boldsymbol{v}_j(\zeta)$, respectively. These quantities are defined through a spectral parameter λ as

$$\begin{aligned} &\lambda_j(\zeta) = \omega_j\zeta, \quad \lambda_j^3(\zeta) = \lambda, \\ &\boldsymbol{v}_j(\zeta) = \begin{pmatrix} 1 \\ \lambda_j(\zeta) \\ \lambda_j^2(\zeta) \end{pmatrix}, \quad \tilde{\boldsymbol{v}}_j(\zeta) = \begin{pmatrix} \lambda_j^2(\zeta) & \lambda_j(\zeta) & 1 \end{pmatrix}, \end{aligned} \tag{9.1.6}$$

where $\omega_j = e^{2\pi i(j-1)/3}$ are the cube roots of 1 ($j = 1, 2, 3$). Obviously the $\lambda_j(\zeta)$ are distinct and they and $\tilde{\boldsymbol{v}}_j(\zeta)$ and $\boldsymbol{v}_j(\zeta)$ are analytic throughout the complex ζ-plane.

The solution of the system of the linear equations (9.1.4) has been obtained by Caudrey [132, 140] in terms of Jost functions $\boldsymbol{\phi}_j(X, \zeta)$ which have the asymptotic behaviour

$$\boldsymbol{\Phi}_j(X, \zeta) := \exp\{-\lambda_j(\zeta)X\}\boldsymbol{\phi}_j(X, \zeta) \to \boldsymbol{v}_j(\zeta) \text{ as } X \to -\infty. \tag{9.1.7}$$

Caudrey [132] showed how the Eq. (9.1.4) can be solved by expressing it as a Fredholm integral equation.

The complex ζ-plane is to be divided into regions such that, in the interior of each region, the order of the numbers $\mathrm{Re}(\lambda_i(\zeta))$ is fixed (see Fig. 7.1). As we pass from one region to another this order changes and hence, on a boundary between two regions, $\mathrm{Re}(\lambda_i(\zeta)) = \mathrm{Re}(\lambda_j(\zeta))$ for at least one pair $i \neq j$. The Jost function $\boldsymbol{\phi}_j$ is regular throughout the complex ζ-plane apart from poles and finite singularities on the boundaries between the regions. At any point in the interior of any region of the complex ζ-plane, the solution of Eq. (9.1.4) is obtained by the relation (2.12) from [132]. It is the direct spectral problem.

We will consider only the inverse spectral problem, i.e. from preassigned spectral data we will reconstruct the solution W. The information about the singularities of the Jost functions $\boldsymbol{\phi}_j(X, \zeta)$ reside in the spectral data. First let us consider the poles. It is assumed that a pole $\zeta_i^{(k)}$ of $\boldsymbol{\phi}_i(X, \zeta)$ does not coincide with a pole of $\boldsymbol{\phi}_j(X, \zeta)$, $j \neq i$ and does not lie on a boundary between two regions. Note that, for $\boldsymbol{\phi}_j(X, \zeta_i^{(k)})$, the point $\zeta_i^{(k)}$ lies in the interior of a regular region. We will need the well-known relations for simple poles [132, 140] in order to compare them with new results which will be obtained in Sec. 9.2. As proven in [132] (see Eq. (7.1.18)), the residue of a simple pole can be calculated as

$$\mathrm{Res}\,\boldsymbol{\phi}_i(X, \zeta_i^{(k)}) = \sum_{\substack{j=1 \\ j \neq i}}^{n} \gamma_{ij}^{(k)} \boldsymbol{\phi}_j(X, \zeta_i^{(k)}). \tag{9.1.8}$$

The quantities $\zeta_i^{(k)}$ and $\gamma_{ij}^{(k)}$ constitute the discrete part of the spectral data in the case of simple poles.

In contrast to Chapter 7 (see [132, 140] too) we do not restrict ourselves to simple poles. Indeed, one of the results we will prove in the next section is that the two-multiple poles can be taken into account in the discrete part of the spectral data. As result the relation (9.1.8) is changed.

Let us consider the singularities on the boundaries between regions. The solution of the spectral problem can be facilitated by using various symmetry properties. In view of (9.1.4), we need only consider the first elements of

$$\boldsymbol{\phi}_i(X,\zeta) = \begin{pmatrix} \phi_i(X,\zeta) \\ \phi_i(X,\zeta)_X \\ \phi_i(X,\zeta)_{XX} \end{pmatrix}, \tag{9.1.9}$$

while the symmetry

$$\phi_1(X,\zeta/\omega_1) = \phi_2(X,\zeta/\omega_2) = \phi_3(X,\zeta/\omega_3) \tag{9.1.10}$$

means we need only consider $\phi_1(X,\zeta)$. In our case, for $\phi_1(X,\zeta)$, the complex ζ-plane is divided into four regions by two lines (see Fig. 7.1) given by

$$\begin{aligned} &\text{(i)} \quad \zeta' = \omega_2\xi, \quad \text{where} \quad \mathrm{Re}(\lambda_1(\zeta)) = \mathrm{Re}(\lambda_2(\zeta)), \\ &\text{(ii)} \quad \zeta' = -\omega_3\xi, \quad \text{where} \quad \mathrm{Re}(\lambda_1(\zeta)) = \mathrm{Re}(\lambda_3(\zeta)), \end{aligned} \tag{9.1.11}$$

where ξ is real. The singularity of $\phi_1(X,\zeta)$ can appear only on these boundaries between the regular regions on the ζ-plane and it is characterized by functions $Q_{1j}(\zeta')$ at each fixed $j \neq 1$. We denote the limit of a quantity, as the boundary is approached, by the superfix $\pm$ in according to the sign of $\mathrm{Re}(\lambda_1(\zeta) - \lambda_j(\zeta))$ (see Fig. 7.1).

In [132] (see Eq. (3.14) there) the jump of $\phi_1(X,\zeta)$ on the boundaries is calculated as

$$\phi_1^+(X,\zeta) - \phi_1^-(X,\zeta) = \sum_{j=2}^{3} Q_{1j}(\zeta)\phi_j^-(X,\zeta), \tag{9.1.12}$$

where, from (9.1.11), the sum is over the lines $\zeta' = \omega_2\xi$ and $\zeta' = -\omega_3\xi$ given by

$$\begin{aligned} &\text{(i)} \quad \zeta' = \omega_2\xi, \quad \text{with} \quad Q_{12}^{(1)}(\zeta') \neq 0, \quad Q_{13}^{(1)}(\zeta') \equiv 0, \\ &\text{(ii)} \quad \zeta' = -\omega_3\xi, \quad \text{with} \quad Q_{12}^{(2)}(\zeta') \equiv 0, \quad Q_{13}^{(2)}(\zeta') \neq 0. \end{aligned} \tag{9.1.13}$$

The quantities $Q_{1j}(\zeta')$ along all the boundaries constitute the continuum part of the spectral data.

The singularity functions $Q_{1j}(\zeta')$ are determined by $W(X,0)$ through the matrix $\mathbf{B}(X,\zeta)$ (7.1.15) (see Eq. (3.13) in [132])

$$Q_{1j}(\zeta) = \frac{1}{\tilde{\boldsymbol{v}}_j(\zeta)\cdot\boldsymbol{v}_j(\zeta)}\tilde{\boldsymbol{v}}_j(\zeta)\cdot\int\limits_{-\infty}^{\infty}\exp[(\lambda_1(\zeta)-\lambda_j(\zeta))z]\mathbf{B}(z,\zeta)\cdot\boldsymbol{\phi}_1^-(X,\zeta)\,dz. \tag{9.1.14}$$

The quantities $Q_{1j}(\zeta')$ along all the boundaries constitute the continuum part of the spectral data.

Thus, for simple poles, the spectral data are [132, 140]

$$S = \{\zeta_1^{(k)},\,\gamma_{1j}^{(k)},\,Q_{1j}(\zeta');j=2,\,3,\,k=1,\,2,\,\ldots,\,m\}. \tag{9.1.15}$$

One of the important features which is to be noted for the IST method is as follows. After the spectral data have been obtained, we need to seek the time-evolution of the spectral data. In Refs. [84, 91, 85, 86, 87] it is proved that for the VPE the T-dependence is revealed as

$$\phi_i(X,T,\zeta) = \exp\left[-\left(3\lambda_i(\zeta)\right)^{-1}T\right]\phi_i(X,0,\zeta),$$

then for spectral data (9.1.15)

$$\begin{aligned}
&\zeta_j^{(k)}(T) = \zeta_j^{(k)}(0),\\
&\gamma_{1j}^{(k)}(T) = \gamma_{1j}^{(k)}(0)\exp\left\{\left[-\left(3\lambda_j(\zeta_1^{(k)})\right)^{-1}+\left(3\lambda_1(\zeta_1^{(k)})\right)^{-1}\right]T\right\},\\
&Q_{1j}(T;\zeta') = Q_{1j}(0;\zeta')\exp\left\{\left[-\left(3\lambda_j(\zeta')\right)^{-1}+\left(3\lambda_1(\zeta')\right)^{-1}\right]T\right\}.
\end{aligned} \tag{9.1.16}$$

The final step in the application of the IST method is to reconstruct the matrix $B(X,T;\zeta)$ and the solution $W(X,T)$ from the spectral data S (9.1.15).

Caudrey has proved that for simple poles the spectral data define $\Phi_1(X,\zeta)$ uniquely in the form (see Eq. (6.20) in [132]))

$$\Phi_1(X,T;\zeta) = 1-\Omega_d(X,T;\zeta)+\Omega_c(X,T;\zeta), \tag{9.1.17}$$

where

$$\Omega_d(X,T;\zeta) \equiv \sum_{k=1}^{K}\sum_{j=2}^{3}\gamma_{1j}^{(k)}(T)\frac{\exp\{[\lambda_j(\zeta_1^{(k)})-\lambda_1(\zeta_1^{(k)})]X\}}{\lambda_1(\zeta_1^{(k)})-\lambda_1(\zeta)} \times\Phi_1(X,T;\omega_j\zeta_1^{(k)}), \tag{9.1.18}$$

$$\Omega_c(X,T;\zeta) \equiv \frac{1}{2\pi\mathrm{i}}\int\sum_{j=2}^{3}Q_{1j}(T;\zeta')\frac{\exp\{[\lambda_j(\zeta')-\lambda_1(\zeta')]X\}}{\zeta'-\zeta} \times\Phi_1^{-}(X,T;\omega_j\zeta')d\zeta'. \tag{9.1.19}$$

Equations (9.1.17)–(9.1.19) contain the spectral data, namely K simple poles with the quantities $\gamma_{1j}^{(k)}$ for the bound state spectrum as well as the functions $Q_{1j}(\zeta')$ given along all the boundaries of regular regions for the continuous spectrum. The integral in (9.1.18) is along all the boundaries (see the dashed lines in Fig. 7.1). The direction of integration is taken so that the side chosen to be $\mathrm{Re}(\lambda_1(\zeta)-\lambda_j(\zeta))<0$ is shown by the arrows in Fig. 7.1 (for the lines (9.1.11), ξ sweeps from $-\infty$ to $+\infty$).

By knowing $\Phi_1(X,T;\zeta)$, we can take extra information into account, namely that the expansion of $\Phi_1(X,T;\zeta)$ as an asymptotic series in $\lambda_1^{-1}(\zeta)$ connects with $W(X,T)$ as follows (cf. Eq. (2.7) in [131]):

$$\Phi_1(X,T;\zeta) = 1-\frac{1}{3\lambda_1(\zeta)}\left[W(X,T)-W(-\infty,T)\right]+O(\lambda_1^{-2}(\zeta)). \tag{9.1.20}$$

Consequently, the solution $W(X,T)$ and the matrix $B(X,T;\zeta)$ can be reconstructed from the spectral data.

In the remaining sections we will study both the multiple poles for the discrete part of the spectral data and the continuum part of the spectral data in special form. Apart from the relations (9.1.18) and (9.1.8), all other formulas are true for the suggested spectral data and will be used subsequently.

9.2 The two-multiple poles

For single poles the formula (9.1.18) are true. Now we take into account the two-multiple poles. Let us consider the additional equation to the spectral equation (9.1.2)

$$\chi_{XXX} + W_{\zeta X}\psi_X + W_X\chi_X - \zeta^3\chi - 3\zeta^2\psi = 0. \tag{9.2.1}$$

For $\chi = \psi_\zeta$ the equation (9.2.1) stems from (9.1.2) by differentiation with respect to ζ. For convenience, the spectral parameter λ is written as $\lambda = \zeta^3$ by virtue of (9.1.6).

The matrix form of the system of equations (9.1.2) and (9.2.1) is as (9.1.4) with

$$\boldsymbol{\psi} = \begin{pmatrix} \psi \\ \psi_X \\ \psi_{XX} \\ \chi \\ \chi_X \\ \chi_{XX} \end{pmatrix}, \quad \mathbf{A} = \begin{pmatrix} 0 & 1 & 0 & 0 & 0 & 0 \\ 0 & 0 & 1 & 0 & 0 & 0 \\ \zeta^3 & 0 & 0 & 0 & 0 & 0 \\ 0 & 0 & 0 & 0 & 1 & 0 \\ 0 & 0 & 0 & 0 & 0 & 1 \\ 3\zeta^2 & 0 & 0 & \zeta^3 & 0 & 0 \end{pmatrix},$$

$$\mathbf{B} = \begin{pmatrix} 0 & 0 & 0 & 0 & 0 & 0 \\ 0 & 0 & 0 & 0 & 0 & 0 \\ 0 & -W_X & 0 & 0 & 0 & 0 \\ 0 & 0 & 0 & 0 & 0 & 0 \\ 0 & 0 & 0 & 0 & 0 & 0 \\ 0 & -W_{\zeta X} & 0 & 0 & -W_X & 0 \end{pmatrix}. \tag{9.2.2}$$

The matrix $\mathbf{A}$ has three pairs of 2-multiple eigenvalues and right-eigenvectors

$$\lambda_j(\zeta) = \lambda_{j+3}(\zeta), \quad \lambda_j(\zeta) = \omega_j\zeta, \quad \lambda_j^3(\zeta) = \lambda,$$

$$\boldsymbol{v}_j(\zeta) = \begin{pmatrix} 0 \\ 0 \\ 0 \\ 1 \\ \lambda_j(\zeta) \\ \lambda_j^2(\zeta) \end{pmatrix} = \boldsymbol{v}_{j+3}(\zeta) = \begin{pmatrix} 0 \\ 0 \\ 0 \\ 1 \\ \lambda_{j+3}(\zeta) \\ \lambda_{j+3}^2(\zeta) \end{pmatrix}, \quad j = 1, 2, 3. \tag{9.2.3}$$

It is known [141] that for the system (9.1.2), (9.2.1) at $W = 0$ to every pair of vectors $\boldsymbol{v}_j(\zeta)$, $\boldsymbol{v}_{j+3}(\zeta)$ $(j = 1, 2, 3)$, there corresponds a system of solutions

$$\boldsymbol{\psi}_j = \boldsymbol{v}_j \exp(\lambda_j X), \quad \boldsymbol{\psi}_{j+3} = (\boldsymbol{v}_j + X\boldsymbol{v}_{2j}) \exp(\lambda_j X), \qquad (9.2.4)$$

where (see p. 97 in [141])

$$\mathbf{A}\boldsymbol{v}_j = \lambda_j \boldsymbol{v}_j, \quad \mathbf{A}\boldsymbol{v}_{2j} = \lambda_j \boldsymbol{v}_{2j} + \boldsymbol{v}_j. \qquad (9.2.5)$$

The multiplicity of eigenvalues does not allow us to obtain the fundamental system of solutions for the system (9.1.2), (9.2.1). To avoid this obstacle we introduce the equation

$$\chi_{XXX} + W_{\zeta X}\psi_X + W_X\chi_X - (\zeta+\varepsilon)^3\chi - 3(\zeta+\varepsilon)^2\psi = 0 \quad (9.2.6)$$

instead of equation (9.2.1). The system (9.1.2), (9.2.6) in matrix form has the matrix

$$\mathbf{A} = \begin{pmatrix} 0 & 1 & 0 & 0 & 0 & 0 \\ 0 & 0 & 1 & 0 & 0 & 0 \\ \zeta^3 & 0 & 0 & 0 & 0 & 0 \\ 0 & 0 & 0 & 0 & 1 & 0 \\ 0 & 0 & 0 & 0 & 0 & 1 \\ 3(\zeta+\varepsilon)^2 & 0 & 0 & (\zeta+\varepsilon)^3 & 0 & 0 \end{pmatrix} \qquad (9.2.7)$$

with different eigenvalues and right-eigenvectors, which in the first approximation $O(\varepsilon)$ have the forms

$$\lambda_j(\zeta) = \omega_j\zeta, \quad \lambda_{j+3}(\zeta) = \omega_j(\zeta+\varepsilon),$$

$$\boldsymbol{v}_j(\zeta) = \begin{pmatrix} -\varepsilon \\ -\varepsilon\lambda_j \\ -\varepsilon\lambda_j^2 \\ 1 \\ \lambda_j \\ \lambda_j^2 \end{pmatrix}, \quad \boldsymbol{v}_{j+3}(\zeta) = \begin{pmatrix} 0 \\ 0 \\ 0 \\ 1 \\ \lambda_{j+3} \\ \lambda_{j+3}^2 \end{pmatrix}, \quad j = 1, 2, 3. \qquad (9.2.8)$$

As $\varepsilon \to 0$ the relations (9.2.8) tend to (9.2.3). At $W = 0$ the solutions of the system (9.1.2), (9.2.6) are

$$\boldsymbol{\psi}_j = \boldsymbol{v}_j \exp(\lambda_j X), \quad j = 1, \ldots, 6. \qquad (9.2.9)$$

In the accepted approximation $O(\varepsilon)$, we take $\boldsymbol{\psi}_{j+3} = \boldsymbol{v}_{j+3}(1+\omega_j\varepsilon X)\exp(\lambda_j X)$ (here $j = 1, 2, 3$) then (9.2.9) is in accord with (9.2.4).

Since the eigenvalues (9.2.8) for the matrix **A** (9.2.7) are different, we can state that a fundamental system of solutions for the system of the equations (9.1.2), (9.2.6) exists (here, for the sake of convenience, the variable X is omitted), namely

$$\boldsymbol{\phi}_j(\lambda_j(\zeta)), \quad j = 1, \ldots, 6. \tag{9.2.10}$$

According to [132, 140] we consider the Wronskian

$$Wr = \det\left[\,\boldsymbol{\phi}_1(\lambda_1),\, \boldsymbol{\phi}_2(\lambda_2),\, \ldots\,, \boldsymbol{\phi}_6(\lambda_6)\,\right]. \tag{9.2.11}$$

If the Wronskian Wr is non-zero at least at one point X_0, then it is proved in [141] (see p. 132 there) to be finite and non-zero even when ζ approaches a pole.

Let $\boldsymbol{\phi}_1(\lambda_1(\zeta))$ have poles at $\zeta = \zeta_1^{(k)}$, $(k = 1, 2)$. Then

$$(\zeta - \zeta_1^{(k)})Wr = \det\left[(\zeta - \zeta_1^{(k)})\boldsymbol{\phi}_1(\lambda_1),\, \boldsymbol{\phi}_2(\lambda_2),\, \ldots\,, \boldsymbol{\phi}_6(\lambda_6)\right]$$

and taking the limit $\zeta \to \zeta_1^{(k)}$ we obtain

$$0 = \det\left[\,\mathrm{Res}\,\boldsymbol{\phi}_1(\lambda_1),\, \boldsymbol{\phi}_2(\lambda_2),\, \ldots\,, \boldsymbol{\phi}_6(\lambda_6)\,\right]. \tag{9.2.12}$$

Thus the columns (vectors) are linearly dependent. The dependence on the vector $\boldsymbol{\phi}_4(\lambda_4)$ is omitted, since it has the same poles as $\boldsymbol{\phi}_1(\lambda_1)$ at $\varepsilon \to 0$.

As a result from (9.2.12), we obtain the solution of the spectral equation (9.1.2) for the bound state spectrum

$$\begin{aligned}\Phi_1(X;\zeta) = 1 - \sum_{k=1}^{2}\sum_{j=2}^{3}\Bigg[&\tilde{\gamma}_{1j}^{(k)}\frac{\exp\{[\lambda_j(\zeta_1^{(k)}) - \lambda_1(\zeta_1^{(k)})]X\}}{\lambda_1(\zeta_1^{(k)}) - \lambda_1(\zeta)} \\ &\times\Phi_1(X;\omega_j\zeta_1^{(k)}) \\ &+\tilde{\gamma}_{1j+3}^{(k)}\frac{\exp\{[\lambda_j(\zeta_1^{(k)}+\varepsilon^{(k)}) - \lambda_1(\zeta_1^{(k)}+\varepsilon^{(k)})]X\}}{\lambda_1(\zeta_1^{(k)}+\varepsilon^{(k)}) - \lambda_1(\zeta)} \\ &\times\Phi_1(X;\omega_j(\zeta_1^{(k)}+\varepsilon^{(k)}))\Bigg].\end{aligned} \tag{9.2.13}$$

By expanding the functions depending on $\varepsilon^{(k)}$ in series within accuracy of $O(\varepsilon^{(k)})$, we rewrite the solution

$$\begin{aligned}\Phi_1(X;\zeta) = 1 - \sum_{k=1}^{2}\sum_{j=2}^{3}\Bigg\{ &\gamma_{1j}^{(k)} \frac{\exp\{[\lambda_j(\zeta_1^{(k)}) - \lambda_1(\zeta_1^{(k)})]X\}}{\lambda_1(\zeta_1^{(k)}) - \lambda_1(\zeta)} \\ &\times \Phi_1(X;\omega_j\zeta_1^{(k)}) \\ &+\frac{\partial}{\partial\zeta_1^{(k)}}\Bigg[\gamma_{1j+3}^{(k)} \frac{\exp\{[\lambda_j(\zeta_1^{(k)}) - \lambda_1(\zeta_1^{(k)})]X\}}{\lambda_1(\zeta_1^{(k)}) - \lambda_1(\zeta)} \\ &\times\Phi_1(X;\omega_j\zeta_1^{(k)})\Bigg]\Bigg\},\end{aligned} \tag{9.2.14}$$

where $\gamma_{1j}^{(k)} = \tilde{\gamma}_{1j}^{(k)} + \tilde{\gamma}_{1j+3}^{(k)}$, $\gamma_{1j+3}^{(k)} = \varepsilon^{(k)}\tilde{\gamma}_{1j+3}^{(k)}$. It is important to note that the solution (9.2.14) is independent of $\varepsilon^{(k)}$ now.

The relationship (9.2.14) formally passes into (9.3.9), (9.3.10) with appropriate change of variables. For this reason the reconstruction of the solution W for (9.2.14) is similar to the problem we will consider for the special form of continuum states (9.3.9).

9.3 The inverse spectral problem for a special continuum spectrum

9.3.1 Special form for the continuum part of the spectral data

Now we consider the continuous spectrum of the associated eigenvalue problem (9.1.4), (9.1.5), (9.1.6), i.e. assume that at least some of the functions $Q_{1j}(\zeta')$ are non-zero. At each fixed $j \neq 1$ the functions $Q_{1j}(\zeta')$ characterize the singularity of $\Phi_1(X,\zeta)$. As we have shown, this singularity can appear only on boundaries between the regular regions on the ζ-plane, where the condition $\mathrm{Re}(\lambda_1(\zeta') - \lambda_j(\zeta')) = 0$ defines these boundaries [132]. For the VPE (9.1.1), as we know, the complex ζ-plane is divided into four regions by two lines (9.1.13)

(i) $\zeta' = \omega_2\xi$, with $Q_{12}^{(1)}(\zeta') \neq 0$, $Q_{13}^{(1)}(\zeta') \equiv 0$,

(ii) $\zeta' = -\omega_3\xi$, with $Q_{12}^{(2)}(\zeta') \equiv 0$, $Q_{13}^{(2)}(\zeta') \neq 0$, where ξ is real (see Fig. 7.1) and sweeps from $-\infty$ to $+\infty$.

Recently in [85, 86, 87] we have considered the singularity functions $Q_{1j}(\zeta')$ on the boundaries, on which the Jost function $\phi_1(X,\zeta)$ is singular, in the form ($m = 1, 2, ..., M$) on the line $\zeta' = \omega_2\xi$

$$\begin{aligned} Q_{12}^{(1)}(\zeta') &= -2\pi\mathrm{i} \sum_{m=1}^{M} q_{12}^{(2m-1)} \delta(\zeta' - \zeta'_{2n-1}), \\ Q_{13}^{(1)}(\zeta') &= -2\pi\mathrm{i} \sum_{m=1}^{M} q_{13}^{(2m-1)} \delta(\zeta' - \zeta'_{2n-1}) \equiv 0, \end{aligned} \tag{9.3.1}$$

and on the line $\zeta' = -\omega_3\xi$

$$\begin{aligned} Q_{12}^{(2)}(\zeta') &= -2\pi\mathrm{i} \sum_{m=1}^{M} q_{12}^{(2m)} \delta(\zeta' - \zeta'_{2n}) \equiv 0, \\ Q_{13}^{(2)}(\zeta') &= -2\pi\mathrm{i} \sum_{m=1}^{M} q_{13}^{(2m)} \delta(\zeta' - \zeta'_{2n}). \end{aligned} \tag{9.3.2}$$

Now we extend the functional dependence for $Q_{1j}(\zeta')$. We focus on the step-function as a possible singularity function

$$f(x) = \frac{1}{h}\left(\Theta(x) - \Theta(x-h)\right), \tag{9.3.3}$$

where $\Theta(x)$ is a Heavyside function. Expanding the Heavyside function $\Theta(x-h)$ into a Taylor series in the neighborhood of the point x

$$\Theta(x-h) = \Theta(x) + \sum_{n=1}^{\infty} (-1)^n \frac{h^n}{n!} \Theta^{(n)}(x), \tag{9.3.4}$$

the step-function (9.3.3) can be rewritten in terms of the derivatives $\delta^{(n)}(x) = \Theta^{(n+1)}(x)$ as follows

$$\begin{aligned} f(x) &= \sum_{n=1}^{\infty} (-1)^{n+1} \frac{h^{n-1}}{n!} \Theta^{(n)}(x) = \sum_{n=0}^{\infty} (-1)^n \frac{h^n}{(n+1)!} \delta^{(n)}(x) \\ &= \delta(x) - \tfrac{1}{2} h \delta^{(1)}(x) + \dots \,. \end{aligned} \tag{9.3.5}$$

We restrict our consideration to only two terms of the series (9.3.5) for modelling the singularity functions $Q_{1j}(\zeta')$. In the limit $h \to 0$, the functions $Q_{1j}(\zeta') = \dfrac{-2\pi\mathrm{i}}{h}\, q_{1j}(\Theta(\zeta') - \Theta(\zeta' - h))$ have to be subject to the relations (9.3.1), (9.3.2). Therefore the singularity functions $Q_{1j}(\zeta')$ that we will examine have the following forms ($m = 1$) on the line $\zeta' = \omega_2\xi$:

$$Q_{12}^{(1)}(\zeta') = -2\pi\mathrm{i}\left(q_{12}^{(1)}\delta(\zeta' - \zeta_1') - \tfrac{1}{2}q_{12}^{(1)}h_1\delta^{(1)}(\zeta' - \zeta_1')\right),$$

$$Q_{13}^{(1)}(\zeta') \equiv 0, \quad \text{i.e.} \quad q_{13}^{(1)} \equiv 0, \tag{9.3.6}$$

$$h_1 = h^{(1)},$$

and on the line $\zeta' = -\omega_3\xi$:

$$Q_{12}^{(2)}(\zeta') \equiv 0, \quad \text{i.e.} \quad q_{12}^{(2)} \equiv 0,$$

$$Q_{13}^{(2)}(\zeta') = -2\pi\mathrm{i}\left(q_{13}^{(2)}\delta(\zeta' - \zeta_2') - \tfrac{1}{2}q_{13}^{(2)}h_2\delta^{(1)}(\zeta' - \zeta_2')\right), \tag{9.3.7}$$

$$h_2 = h^{(2)}.$$

Consequently, the spectral data for the continuum spectrum with special singularity functions (9.3.6), (9.3.7) are

$$S = \{\zeta_l',\, q_{1j}^{(l)},\, h_l\,;\, j = 2,\,3,\; l = 1,\,2\}. \tag{9.3.8}$$

9.3.2 Reconstructing the solution

Let us consider the problem of reconstructing the solution $W(X)$ from the spectral data (9.3.8). This will be straightforward if we can find the vectors $\Phi_1(X,T;\zeta)$. Now we study only the special form of the continuum part of the spectral data (9.3.6), (9.3.7), while the variable $\Omega_d(X,T;\zeta)$ given in (9.1.18) is considered to be identically zero. For the singularity functions (9.3.6), (9.3.7) the relationship (9.1.17) with (9.1.19) is reduced to the form (provisionally the time-dependence is not written)

$$\begin{aligned}\Phi_1(X,\zeta) = \;& 1 - \sum_{l=1}^{2}\sum_{j=2}^{3}\Bigg[q_{1j}^{(l)} L_j(X;\zeta_l',\zeta)\Phi_1(X,\omega_j\zeta_l') \qquad (9.3.9)\\ & + \frac{1}{2}\, q_{1j}^{(l)} h_l \left(\frac{\partial}{\partial\zeta'} L_j(X;\zeta',\zeta)\Phi_1(X,\omega_j\zeta')\right)_{\zeta'=\zeta_l'}\Bigg],\end{aligned}$$

where

$$L_j(X;\zeta',\zeta) \equiv \frac{\exp\{[\lambda_j(\zeta') - \lambda_1(\zeta')]X\}}{\zeta' - \zeta}. \tag{9.3.10}$$

We note once again that the relationships (9.2.14) and (9.3.9) are similar.

As was proved in Refs.[85, 86, 87], the singularities appear in pairs

$$\zeta_1' = \omega_2\xi_1, \quad \zeta_2' = -\omega_3\xi_1, \tag{9.3.11}$$

where ξ_1 is a real constant. Moreover

$$\omega_2 q_{12}^{(1)} = q_{13}^{(2)}. \tag{9.3.12}$$

It is evident that from (9.3.11)

$$h_1 = \omega_2 h, \quad h_2 = -\omega_3 h,$$

where h is a real constant.

Here it is convenient to note that the time-evolution of the spectral data appears through (9.1.16) in the form

$$\xi_1 = \text{const}, \quad h = \text{const}, \quad q_{1j}^{(k)}(T) = q_{1j}^{(k)}(0)\exp\left(\frac{1}{\mathrm{i}\sqrt{3}}\frac{T}{\xi_1}\right). \tag{9.3.13}$$

The equation (9.3.9) allows us to define the functions $\Phi_1(X,\zeta)$. Indeed, differentiating this equation (9.3.9) with respect of ζ, and substituting the values $\zeta = \omega_2\zeta_1'$, $\zeta = \omega_3\zeta_2'$ in the left-hand side of these equations, we obtain a system of four linear algebraic equations in the unknowns

$$\Phi_1(X,\omega_2\zeta_1'), \Phi_1(X,\omega_3\zeta_2'),$$

$$\left.\frac{\partial}{\omega_2\partial\zeta}\Phi_1(X,\omega_2\zeta)\right|_{\zeta=\zeta_1'}, \left.\frac{\partial}{\omega_3\partial\zeta}\Phi_1(X,\omega_3\zeta)\right|_{\zeta=\zeta_2'}.$$

Hence, we could take the function $\Phi_1(X,\zeta)$ from Eq. (9.3.9).

However, there is a more direct method, in which there is no need to obtain the variables $\Phi_1(X,\omega_2\zeta_1')$, $\Phi_1(X,\omega_3\zeta_2')$ explicitly. It turns out that we need to calculate only a determinant of some matrix. This approach is similar to the method referred to in [132, 140, 84, 85, 86, 87]. It is convenient to use new variables introduced by the definition

$$\Psi_l(X;\zeta_l') = \sum_{j=2}^{3} q_{1j}^{(l)} \exp(\lambda_j(\zeta_l')X)\Phi_1(X,\omega_j\zeta_l'), \quad l=1,\,2, \tag{9.3.14}$$

i.e.

$$\Psi_1(X;\zeta_1') = q_{12}^{(1)} \exp(\lambda_2(\zeta_1')X)\Phi_1(X,\omega_2\zeta_1'),$$

$$\Psi_2(X;\zeta_2') = q_{13}^{(2)} \exp(\lambda_3(\zeta_2')X)\Phi_1(X,\omega_3\zeta_2').$$

We may rewrite the relationship (9.3.9) as

$$\begin{aligned}\Phi_1(X;\zeta) = \quad & 1 - \sum_{l=1}^{2} \frac{\exp(-\lambda_1(\zeta_l')X)}{\zeta_l' - \zeta}\Psi_l(X;\zeta_l') \\ & + \sum_{l=1}^{2} \frac{1}{2} h_l \frac{\partial}{\partial \zeta_l'} \left(\frac{\exp(-\lambda_1(\zeta_l')X)}{\zeta_l' - \zeta}\Psi_l(X;\zeta_l') \right).\end{aligned} \tag{9.3.15}$$

Here we introduce the notations

$$\begin{aligned} L(X;\zeta,\zeta_l') & \equiv \frac{\exp\{[\lambda_1(\zeta) - \lambda_1(\zeta_l')]X\}}{\zeta_l' - \zeta} \\ & = - \int_X \exp\{[\lambda_1(\zeta) - \lambda_1(\zeta_l')]X'\} dX', \end{aligned} \tag{9.3.16}$$

and then

$$\frac{\partial}{\partial \zeta_l'} L(X;\zeta,\zeta_l') = \int_X X' \exp\{[\lambda_1(\zeta) - \lambda_1(\zeta_l')]X'\} dX', \tag{9.3.17}$$

$$\frac{\partial^2}{\partial \zeta \partial \zeta_l'} L(X;\zeta,\zeta_l') = \int_X X'^2 \exp\{[\lambda_1(\zeta) - \lambda_1(\zeta_l')]X'\} dX'. \tag{9.3.18}$$

Taking into account (9.1.20), namely

$$\Phi_1(X,\zeta) = 1 - \frac{1}{3\lambda_1(\zeta)}\left[W(X) - W(-\infty)\right] + O(\lambda_1^{-2}(\zeta)),$$

and (9.3.14) and (9.3.15), the following relationship may be found

$$-\frac{1}{3}\left[W(X)-W(-\infty)\right] = \sum_{l=1}^{2}\Big[\exp(-\lambda_1(\zeta_l')X)\Psi_l(X;\zeta_l') - \frac{1}{2}h_l\frac{\partial}{\partial\zeta_l'}\exp(-\lambda_1(\zeta_l')X)\Psi_l(X;\zeta_l')\Big]. \tag{9.3.19}$$

Eq. (9.3.15) with (9.3.14) in notations (9.3.16)–(9.3.18) can be rewritten as follows:

$$\exp(\lambda_1\ (\zeta)X)\Phi_1(X;\zeta) = \exp(\lambda_1(\zeta)X) - \sum_{l=1}^{2} L(X;\zeta,\zeta_l')\Psi_l(X;\zeta_l') + \sum_{l=1}^{2}\frac{1}{2}h_l\Psi_l(X;\zeta_l')\int_X X'\exp\{[\lambda_1(\zeta)-\lambda_1(\zeta_l')]X'\}dX' + \sum_{l=1}^{2}\frac{1}{2}h_l L(X;\zeta,\zeta_l')\frac{\partial}{\partial\zeta_l'}\Psi_l(X;\zeta_l'). \tag{9.3.20}$$

In contrast to the standard procedure, here it is necessary to take into account the time-evolution for $q_{1j}^{(k)}(T)$ (9.3.13). Differentiating Eq. (9.3.20) with respect to ζ, and substituting the values $\zeta=\omega_2\zeta_1'$, $\zeta=\omega_3\zeta_2'$ in the left-hand side of these equations, we obtain a system of four linear algebraic equations in the unknowns $\Psi_l(X;\zeta_l')$, $\dfrac{\partial}{\partial\zeta_l'}\Psi_l(X;\zeta_l')$ for $l=1$, 2. The matrix form of this system of equations is

$$\mathbf{M\Psi}=\mathbf{b}, \tag{9.3.21}$$

where

$$\mathbf{\Psi}=\begin{pmatrix}\Psi_1(X;\zeta_1')\\ \Psi_2(X;\zeta_2')\\ \omega_3\dfrac{\partial}{\partial\zeta_1'}\Psi_1(X;\zeta_1')\\ \omega_2\dfrac{\partial}{\partial\zeta_2'}\Psi_2(X;\zeta_2')\end{pmatrix},\quad \mathbf{b}=\begin{pmatrix}q_{12}^{(1)}\exp(\omega_2\zeta_1'X)\\ q_{13}^{(2)}\exp(\omega_3\zeta_2'X)\\ q_{12}^{(1)}X\exp(\omega_2\zeta_1'X)\\ q_{13}^{(2)}X\exp(\omega_3\zeta_2'X)\end{pmatrix}. \tag{9.3.22}$$

The elements of matrix $\mathbf{M}$ are

$$
\begin{aligned}
M_{11} &= 1 - q_{12}^{(1)} \frac{\exp(-\mathrm{i}\sqrt{3}\xi_1 X)}{-\mathrm{i}\sqrt{3}\xi_1} - \frac{1}{2}\, q_{12}^{(1)} h_1 \int\limits_X X' \exp(-\mathrm{i}\sqrt{3}\xi_1 X')dX', \\
M_{12} &= -q_{12}^{(1)} \frac{\exp(2\omega_3\xi_1 X)}{2\omega_3\xi_1} - \frac{1}{2}\, q_{12}^{(1)} h_2 \int\limits_X X' \exp(2\omega_3\xi_1 X')dX', \\
M_{13} &= \frac{1}{2}\, q_{12}^{(1)} \omega_2 h_1 \frac{\exp(-\mathrm{i}\sqrt{3}\xi_1 X)}{-\mathrm{i}\sqrt{3}\xi_1}, \\
M_{14} &= \frac{1}{2}\, q_{12}^{(1)} \omega_3 h_2 \frac{\exp(2\omega_3\xi_1 X)}{2\omega_3\xi_1}, \\
M_{21} &= -q_{13}^{(2)} \frac{\exp(-2\omega_2\xi_1 X)}{-2\omega_2\xi_1} - \frac{1}{2}\, q_{13}^{(2)} h_1 \int\limits_X X' \exp(-2\omega_2\xi_1 X')dX', \\
M_{22} &= 1 - q_{13}^{(2)} \frac{\exp(-\mathrm{i}\sqrt{3}\xi_1 X)}{-\mathrm{i}\sqrt{3}\xi_1} - \frac{1}{2}\, q_{13}^{(2)} h_2 \int\limits_X X' \exp(-\mathrm{i}\sqrt{3}\xi_1 X')dX', \\
M_{23} &= \frac{1}{2}\, q_{13}^{(2)} \omega_2 h_1 \frac{\exp(-2\omega_2\xi_1 X)}{-2\omega_2\xi_1}, \\
M_{24} &= \frac{1}{2}\, q_{13}^{(2)} \omega_3 h_2 \frac{\exp(-\mathrm{i}\sqrt{3}\xi_1 X)}{-\mathrm{i}\sqrt{3}\xi_1}, \qquad (9.3.23) \\
M_{31} &= -q_{12}^{(1)} \int\limits_X X' \exp(-\mathrm{i}\sqrt{3}\xi_1 X')dX' \\
&\quad - \frac{1}{2}\, q_{12}^{(1)} h_1 \int\limits_X X'^2 \exp(-\mathrm{i}\sqrt{3}\xi_1 X')dX' + \frac{T}{\mathrm{i}\sqrt{3}\omega_3\xi_1^2}, \\
M_{32} &= -q_{12}^{(1)} \int\limits_X X' \exp(2\omega_3\xi_1 X')dX' \\
&\quad - \frac{1}{2}\, q_{12}^{(1)} h_2 \int\limits_X X'^2 \exp(2\omega_3\xi_1 X')dX', \\
M_{33} &= 1 + \frac{1}{2}\, q_{12}^{(1)} \omega_2 h_1 \int\limits_X X' \exp(-\mathrm{i}\sqrt{3}\xi_1 X')dX',
\end{aligned}
$$

$$M_{34} = \frac{1}{2}\, q_{12}^{(1)} \omega_3 h_2 \int\limits_X X' \exp(2\omega_3 \xi_1 X') dX',$$

$$M_{41} = -q_{13}^{(2)} \int\limits_X X' \exp(-2\omega_2 \xi_1 X') dX'$$
$$-\frac{1}{2}\, q_{13}^{(2)} h_1 \int\limits_X X'^2 \exp(-2\omega_2 \xi_1 X') dX',$$

$$M_{42} = -q_{13}^{(2)} \int\limits_X X' \exp(-\mathrm{i}\sqrt{3}\xi_1 X') dX'$$
$$-\frac{1}{2}\, q_{13}^{(2)} h_2 \int\limits_X X'^2 \exp(-\mathrm{i}\sqrt{3}\xi_1 X') dX' - \frac{T}{\mathrm{i}\sqrt{3}\omega_2 \xi_1^2},$$

$$M_{43} = \frac{1}{2}\, q_{13}^{(2)} \omega_2 h_1 \int\limits_X X' \exp(-2\omega_2 \xi_1 X') dX',$$

$$M_{44} = 1 + \frac{1}{2}\, q_{13}^{(2)} \omega_3 h_2 \int\limits_X X' \exp(-\mathrm{i}\sqrt{3}\xi_1 X') dX'.$$

Note that the time-dependence in the matrix elements appears both through $q_{1j}^{(k)}$ and, in contrast to the standard procedure, through the last terms in M_{31} and M_{42} which appear because $\dfrac{\partial q_{1j}^{(k)}}{\partial \xi_1} \neq 0$.

Since for any column j of the matrix $\mathbf{M}$ we have

$$\exp(\omega_k \xi_1 X) \frac{\partial}{\partial X} M_{ij} = b_i, \quad k = \left\{ \begin{array}{ll} 2, & \text{if } \; i = 2n+1 \\ 3, & \text{if } \; i = 2n+2 \end{array} \right. ,$$

the sum for (9.3.19) is

$$\sum_{l=1}^{2} \left[\exp(-\zeta_l X) \Psi_l(X;\zeta_l) - \frac{1}{2}\, h_l \frac{\partial}{\partial \zeta_l} \exp(-\zeta_l X) \Psi_l(X;\zeta_l) \right]$$
$$= \frac{1}{\det \mathbf{M}} \frac{\partial \det \mathbf{M}}{\partial X} .$$

Finally, from the relation (9.3.19), the following key relationship may be obtained

$$W(X) - W(-\infty) = 3 \frac{\partial}{\partial X} \ln(\det \mathbf{M}(X)). \tag{9.3.24}$$

9.3.3 Calculating the determinant of the matrix M

We will prove that the determinant of the matrix **M** is given by

$$\det \mathbf{M} = \left[1 + \left(s_1 + \mathrm{i} r_1 \left\{X - \frac{T}{3\xi_1^2}\right\}\right) \exp(\theta_1) + p_1 \exp(2\theta_1)\right]^2, \tag{9.3.25}$$

where

$$s_1 = c_1 \left(1 + \frac{h}{2\xi_1}\right), \quad r_1 = \frac{\sqrt{3}}{2} h c_1, \quad p_1 = -\frac{h^2 c_1^2}{3 \cdot 2^4 \xi_1^2}, \tag{9.3.26}$$

$$c_1 = \frac{\beta_1}{-\mathrm{i}2\sqrt{3}\xi_1}, \quad \theta_1 = -\mathrm{i}\sqrt{3}\xi_1 X + \frac{T}{\mathrm{i}\sqrt{3}\xi_1}.$$

Since the singularities occur in pairs, $\det \mathbf{M}$ is to be a perfect square for some auxiliary function F. This statement is not proved directly. However, numerical calculations using the software *Maple* showed that the matrix **M** has two pairs of equal eigenvalues $\lambda_i^{(M)}$ $(i = 1, \ldots, 4)$, i.e. $\lambda_1^{(M)} = \lambda_2^{(M)}$, $\lambda_3^{(M)} = \lambda_4^{(M)}$. It is known that the coefficient in $O(\lambda^2)$ in the eigenfunction of the $[4 \times 4]$ matrix is written

$$\sum_{\substack{i,j=1\\ i<j}}^{4} \det \begin{pmatrix} M_{ii} & M_{ij} \\ M_{ji} & M_{jj} \end{pmatrix}.$$

On the other hand, under conditions $\lambda_1^{(M)} = \lambda_2^{(M)}$, $\lambda_3^{(M)} = \lambda_4^{(M)}$ this coefficient is equal to $2\lambda_1^{(M)}\lambda_3^{(M)} + \left(\lambda_1^{(M)} + \lambda_3^{(M)}\right)^2$. Thus, we have the relationship

$$\sum_{\substack{i,j=1\\ i<j}}^{4} \det \begin{pmatrix} M_{ii} & M_{ij} \\ M_{ji} & M_{jj} \end{pmatrix} = 2\lambda_1^{(M)}\lambda_3^{(M)} + \left(\lambda_1^{(M)} + \lambda_3^{(M)}\right)^2. \tag{9.3.27}$$

In as much as $\mathrm{Tr}\,\mathbf{M} = \sum_{i=1}^{4} M_{ii} = 2\left(\lambda_1^{(M)} + \lambda_3^{(M)}\right)$, and $\det \mathbf{M} = \left(\lambda_1^{(M)}\lambda_3^{(M)}\right)^2$, the relationship (9.3.27) enables us to find the auxiliary function $F = \sqrt{\det \mathbf{M}}$ as follows:

$$F(X) = \sqrt{\det \mathbf{M}} = \frac{1}{4}\sum_{\substack{i,j=1\\ i<j}}^{4} M_{ii}M_{jj} - \frac{1}{2}\sum_{\substack{i,j=1\\ i<j}}^{4} M_{ij}M_{ji} - \frac{1}{8}\sum_{i=1}^{4} M_{ii}^2. \tag{9.3.28}$$

Omitting the cumbersome calculation, we finally obtain the relation (9.3.25).

There are three constants, namely ξ_1, h which are real, and β_1 which could be complex in the general case.

The substitution of the relation (9.3.28) into (9.3.24) and the taking into account of the T–evolution of the spectral data for the VPE [84] (see also (9.1.16)) allows one to find the solution for the special continuum spectrum (9.3.6), (9.3.7) as

$$W(X,T) - W(-\infty,T) = 6\frac{\partial}{\partial X}\ln(F(X,T)). \tag{9.3.29}$$

The problem of selecting the real solution from the complex relation (9.3.29) is open for study.

9.4 The solution for discrete spectral data with two-multiple poles

The results for the continuum part of the spectral data obtained in Sec. 9.3.2 and Sec. 9.3.3 can be reduced to the bound state spectrum since the relationships (9.2.14) and (9.3.9) are similar to each other. The formal replacements

$$h \to \mathrm{i}h, \quad \xi_1 \to \mathrm{i}\xi_1 \tag{9.4.1}$$

lead to the solution (9.3.24) of the VPE for the discrete spectrum with two-multiple poles (9.2.14), namely

$$W(X,T) - W(-\infty,T) = 6\frac{\partial}{\partial X}\ln(F(X,T)) \tag{9.4.2}$$

with auxiliary function

$$F(X,T) = 1+\left(s_2 + r_2\left\{X + \frac{T}{3\xi_1^2}\right\}\right)\exp(\theta_2)+p_2\exp(2\theta_2)\,, \quad (9.4.3)$$

$$s_2 = c_2\left(1+\frac{h}{2\xi_1}\right), \quad r_2 = -\frac{\sqrt{3}}{2}\,hc_2\,, \quad p_2 = -\frac{h^2c_2^2}{3\cdot 2^4\xi_1^2}\,,$$

$$c_2 = \frac{\beta_1}{2\sqrt{3}\xi_1}\,, \quad \theta_2 = \sqrt{3}\xi_1 X - \frac{T}{\sqrt{3}\xi_1}\,.$$

The constants ξ_1, h are real. There is one arbitrary constant β_1. It is to be real for a real solution.

Note that the auxiliary function F is associated with the τ-function (see, for example, [29, 144, 145]).

By taking into account the transformation (9.4.1), we can apply all mathematical manipulations stated in Sec. 9.3.2 and Sec. 9.3.3 to the discrete part of the spectral data.

Since $p_2 < 0$ for arbitrary real β_1, we have $\lim\limits_{X\to-\infty} F = 1$, and $\lim\limits_{X\to+\infty} F = -\infty$, hence there is X_r such that $F(X_r) = 0$. Thus the real solution (9.4.2) with (9.4.3) is a singular function.

If we determine the value β_1 as an imaginary one, the solutions will be smooth but complex. The selection of the real solutions from complex ones is an open problem.

9.5 Two-multiple poles and a single pole

Now we consider the interaction of a soliton with a wave which is associated with a two-multiple pole. Let the soliton be defined by a single pole with value ξ_3. This soliton has the values $c_3 = \dfrac{\beta_1}{2\sqrt{3}\xi_3}$, $\theta_3 = \sqrt{3}\xi_3 X - \dfrac{T}{\sqrt{3}\xi_3}$.

For convenience, we rewrite the relation (9.4.3) in an alternative form, with the auxiliary function denoted by index $2p$, i.e. $F_{2p}(X,T) \equiv F(X,T)$, namely

$$F_{2p}(X,T) = 1 + c_2\,(1+gh)\exp(\theta_2) + p_2\exp(2\theta_2)\,, \qquad (9.5.1)$$

where $g = \dfrac{1}{2\xi_1} - \dfrac{\sqrt{3}}{2}\left(X + \dfrac{T}{3\xi_1^2}\right)$. Adding this wave to the soliton, we can obtain the solution (9.4.2) by means of the auxiliary function in the form

$$
\begin{aligned}
F_{2ps}(X,T) = \; & 1 + c_2\,(1+gh)\exp(\theta_2) + c_3\exp(\theta_3) + \\
& p_2\exp(2\theta_2) + b_{13}[1+(g+g_3)h]c_2c_3\exp(\theta_2)\exp(\theta_3) + \\
& p_2 b_{13}^2 c_3\exp(2\theta_2)\exp(\theta_3),
\end{aligned}
\tag{9.5.2}
$$

where

$$
g_3 = -\frac{1}{2\xi_3}\frac{b_{13p}}{b_{13}}, \quad b_{13} = \frac{(y-1)^3}{(y+1)^3}\frac{y^3+1}{y^3-1}, \quad y = \frac{\xi_1}{\xi_3}, \quad b_{13p} = \frac{d\,b_{13}}{d\,y}.
$$

Thus, we have obtained the solution associated with the interaction of a soliton and a wave that is generated by a two-multiple pole in the discrete spectrum.

9.6 Conclusion

Using the VPE as an example, we have shown how, in the IST method, to take into account the two-multiple poles, among single poles, in the discrete part of the spectral data. The special line spectrum of continuum states in the IST method, for which the mathematical procedure is similar to that for the discrete spectrum for two-multiple poles, is considered as well. New solutions are obtained and verified using direct substitution into the initial equation by Maple software. The account of the time-dependence is different from the standard procedure.

The important problem which remains is finding the connection between the Lax pairs for the VPE and the VE, and can be a matter for scientific enquiry in the future. This problem is difficult because the solutions of the VPE are single-functions, while the loop-like so-lutions of the VE can usually be expressed in parametric form only.

Bibliography

[1] Clarke, J.E. (1984) Lectures on plane waves in reacting gases. *Annals of Physics*, **9**, 211–306.

[2] Landau, L.D. and Lifshitz, E.M. (1988) *Fluids Mechanics*, Pergamon, New York.

[3] Danylenko, V.A., Sorokina, V.V. and Vladimirov, V.A. (1993) On the governing equations in relaxing models and self-similar quasiperiodic solutions. *J. Phys. A: Math. Gen.*, **26**, 7125–7135.

[4] Yasnikov, G.P. and Belousov, V.S. (1978) Effective thermodynamic functions of a gas with solid particles. *Journal of Engineering Physics*, **34**, 734–737. (In Russian)

[5] Mandelshtam, L.I. and Leonovich, M.A. (1937) To the theory of a sound attenuation in liquids. *Journal of Experimental and Theoretical Physics*, **3**, 438–449. (In Russian)

[6] Nakoryakov, V.E., Pokusaev, B.G. and Shreber, I.R. (1983) *Wave Propagation in Gaseous and Vapor-Liquid Media*, Institute of Thermophysics Press, Novosibirsk. (In Russian)

[7] Lyakhov, G.M. (1982) *Waves in Soils and Porous Multicomponent Media*, Nauka, Moscow. (In Russian)

[8] Zener, C. (1948) *Elasticity and Anelasticity of Metals*, The University of Chicago Press, Chicago.

[9] Vakhnenko, V.O., Vakhnenko, O.O., TenCate, J.A. and Shankland, T.J. (2007) Modeling of stress-strain dependences for Berea sandstone under quasistatic loading. *Phys. Rev. B*, **76**, 184108(8).

[10] Biot, M.A. (1956) Theory of propagation of elastic waves in a fluid-saturated solid. I. Low-frequency range. *Journal of the Acoustical Society of America*, **28**, 168–178.

[11] Vakhnenko, V.O. (1999) High-frequency soliton-like waves in a relaxing medium. *J. Math. Phys.*, **40**, 2011–2020.

[12] Vakhnenko, V.A. (1991) *Periodic Short-Wave Perturbations in a Relaxing Medium*, Institute of Geophysics, The Ukrainian Academy of Science, Kiev, Ukraine. (In Russian)

[13] Vakhnenko, V.O. (1997) The existence of loop-like solutions of a model evolution equation. *Ukrainian Journal of Physics*, **42**, 104–110. (In Ukrainian)

[14] Vakhnenko, V.O. (2015) *Wave Dynamics of the Structured Media with Relaxation*, Lap Lambert Academic Publishing, OmniScriptum GmbH and Co. KG, Saarbrucken, Germany.

[15] Mitropolsky, Yu.A., Samoilenko, A.M. and Martinyuk, D.I. (1993) *Systems of Evolution Equations with Periodic and Quasiperiodic Coefficients*, Kluwer Academic, Dordrecht.

[16] Nayfeh, A.H. (1973) *Perturbation methods*, Wiley-Interscience, New York.

[17] Vakhnenko, V.A. (1999) Increase of nonlinear effect in a medium with microstructure. *Acoustical Physics*, **45**, 151–157.

[18] Vakhnenko, V.A., Danilenko, V.A. and Kulich, V.V. (1996) Asymptotic justification of the Lyakhov model for multicomponent media. *Combustion Explosion and Shock Waves*, **32**(2), 176–180.

[19] Vakhnenko, V.A., Danilenko, V.A. and Michtchenko, A.V. (1999) An asymptotic averaged model of nonlinear long waves propagation in media with a regular structure. *Int. J. Non-Linear Mech.*, **34**, 643–654.

[20] Vakhnenko, V.A., Danilenko, V.A. and Kulich, V.V. (1993) Averaged description of wave processes in geophysical medium. *Geophysics J. (Ukraine)*, **15**(6), 66–71.

[21] Vakhnenko, V.A., Danilenko, V.A. and Kulich, V.V. (1994) Averaged description of shock-wave processes in periodic media. *Sov. J. Chem. Phys.*, **12**(3), 534–546.

[22] Vakhnenko, V.O. (2014) Blast waves in multi-component medium with thermal relaxation. *Natural Science*, **6**, 1055–1092.

[23] Kuetche, V.K., Youssoufa, S. and Kofane, T.C. (2014) Phase portraits analysis of a barothropic system: The initial value problem. *J. Math. Phys.*, **55**, 052702.

[24] Kuetche, V.K. (2015) Barothropic relaxing media under pressure perturbations: Nonlinear dynamics. *Dynamics of Atmospheres and Oceans*, **72**, 21–37.

[25] Ablowitz, M. and Segur, H. (1981) *Solitons and the Inverse Scattering Transform*, SIAM, Philadelphia.

[26] Gabov, S.A. (1988) *Introduction to the Theory of Nonlinear Waves*, MGU Press, Moscow. (In Russian)

[27] Dodd, R.K., Eilbeck, J.C., Gibbon, J.D. and Morris, H.C. (1984) *Solitons and Nonlinear Wave Equations*, Academic Press, London.

[28] Novikov, S.P., Manakov, S.V., Pitaevskii, L.P. and Zakharov, V.E. (1984) *Theory of Solitons: The Inverse Scattering Method*, Plenum Publishing Corporation, New York-London.

[29] Newell, A.C. (1985) *Solitons in Mathematics and Physics*, SIAM, Philadelphia.

[30] Su, C.H. and Gardner, C.S. (1969) Korteweg–de Vries equation and generalizations. III. Derivation of Korteweg–de Vries equation and Burgers equation. *J. Math. Phys.*, **10**, 536–539.

[31] Gardner, C.S. and Morikawa, G.K. (1960) Similarity in the asymptotic behaviour of collision free hydromagnetic wave and water waves. Tech. report. *New York University Report NYU-9082, Courant Inst. Math. Scien.*

[32] Washimi, H. and Taniuti, T. (1966) Propagation of ion-acoustic solitary waves of small amplitude. *Phys. Rev. Lett.*, **17**, 996–998.

[33] Faddeev, L.D. and Takhtajan, L.A. (1987) *Hamiltonian Methods in the Theory of Solitons*, Springer-Verlag, Berlin, Heidelberg and New York.

[34] Gardner, C.S., Greene, J.M., Kruskal, M.D. and Miura, R.M. (1967) Method for solving the Korteweg–de Vries equation. *Phys. Rev. Lett.*, **19**, 1095–1097.

[35] Hirota, R. (1980) Direct methods in soliton theory, in *Solitons*, eds. R.K. Bullough and P.J. Caudrey, Springer, New York, Berlin, pp. 157–176.

[36] Hirota, R. (2004) *The Direct Method in Soliton Theory*, Cambridge University Press, Cambridge.

[37] Wazwaz, A.M. (2009) *Partial Differential Equations and Solitary Waves Theory*, Higher Education Press and Springer-Verlag, Beijing and Berlin.

[38] Vladimirov, V.S. (1971) *Equations of Mathematical Physics*, Dekker, New York.

[39] Malykh, K.V. and Ogorodnikov, I.L. (1977) On application of the Klein–Gordon equation for describing the structure of impulse of the compression in a liquid with the bubbles of gas, in *Dynamics of Continuous Media*, Hydrodynamics Institute Press, Novosibirsk, pp. 140–148. (In Russian)

[40] Whitham, G.B. (1974) *Linear and Nonlinear Waves*, Wiley-Interscience, New York.

[41] Vakhnenko, V.A. (1992) Solitons in a nonlinear model medium. *J. Phys. A: Math. Gen.*, **26**(15), 4181–4187.

[42] Parkes, E.J. (1993) The stability of solutions of Vakhnenko's equation. *J. Phys. A: Math. Gen.*, **26**, 6469–6475.

[43] Ostrovsky, L.A. (1978) Nonlinear internal waves in a rotating ocean. *Okeanologia*, **18**, 181–191.

[44] Byrd, P.F. and Friedman, M.D. (1971) *Handbook of Elliptic Integrals for Engineers and Scientists*, Berlin: Springer.

[45] Takahashi, D. and Satsuma, J. (1988) Explicit solutions of magma equation. *J. Phys. Soc. Japan*, **57**, 417–421.

[46] Infeld, E. and Rowlands, G. (1990) *Nonlinear Waves, Solitons and Chaos*, Cambridge, Cambridge University Press.

[47] Murawski, K. and Storer, R.G. (1989) Stability of envelope waves. *Wave Motion*, **11**, 309–325.

[48] Infeld, E., Ziemkiewicz, J. and Rowlands, G. (1987) The two-dimensional stability of periodic and solitary wavetrains on a water surface over arbitrary depth. *Phys. Fluids*, **30**, 2330–2338.

[49] Infeld, E. (1985) Self-focusing of nonlinear ion-acoustic waves and solitons in magnetized plasmas. *J. Plasma Phys.*, **33**, 171–182.

[50] Das K.P., Sluijter F.W. and Verheest F. (1992) Stability of ion-acoustic cnoidal and solitary waves in a magnetized plasma. *Physica Scripta*, **45**, 358–363.

[51] Konno, K., Ichikawa, Y.H. and Wadati, M. (1981) A loop soliton propagating along a stretched rope. *J. Phys. Soc. Japan*, **50**, 1025–1026.

[52] Hasimoto, H. (1972) A soliton on a vortex filament. *J. Fluid Mech.*, **51**, 477–485.

[53] Lamb, G.L. Jr. (1977) Solitons on moving space curves. *J. Math. Phys.*, **18**, 1654–1661.

[54] Burgers, J.M. (1948) A mathematical model illustrating the theory of turbulence. *Advances in Applied Mechanics*, (eds. Richard von Mises, Theodore von Kármán), Academic Press Inc., New York, N.Y. pp. 171–199.

[55] Rajagopal, K.R. and Tao, L. (1995) *Mechanics of Mixtures*, World Scientific Publishing, Singapore.

[56] Struminskii, V.V. (1980) *Mechanics and Technical Progress*, Nauka Press, Moscow. (In Russian)

[57] Nigmatulin, R.I. (1987) *Dynamics of Multiphase Media*, Vol. I and II, Nauka Press, Moscow. (In Russian)

[58] Rudyak, V.N. (1987) *The Statistical Theory of Dissipative Processes in Gases and Liquids*, Nauka, Moscow. (In Russian)

[59] Mikhailov, A.V. and Novikov, V.S. (2002) Perturbative symmetry approach. *J. Phys. A: Math. Gen.*, **35**, 4775–4790.

[60] Guo, B. and Liu, Z. (2004) Periodic cusp wave solutions and single-solitons for the *b*-equation. *Chaos, Solitons and Fractals*, **23**, 1451–1463.

[61] Holm, D.D. and Hone, A.N.W. (2005) A class of equations with peakon and pulson solutions (with an appendix by Harry Braden and John Byatt-Smith). *J. Nonlinear Math. Phys.*, **12** (Suppl. 1), 380–394.

[62] Degasperis, A., Holm, D.D. and Hone, A.N.W. (2002) A new integrable equation with peakon solutions. *Theor. Math. Phys.*, **133**, 1461–1472.

[63] Camassa, R. and Holm, D.D. (1993) An integrable shallow water equation with peaked solitons. *Phys. Rev. Lett.*, **71**, 1661–1664.

[64] Degasperis, A. and Procesi, M. (1999) Asymtotic integrability, in *Symmetry and Perturbation Theory*, eds. A. Degasperis and G. Gaeta, World Scientific, 23–37.

[65] Chen, S., Foias, C., Holm, D., Olson, E., Titi, E.S. and Wynne, S. (1999) A connection between Camassa–Holm equations and turbulent flows in channels and pipes. *Phys. Fluids* **11**, 2343–2353.

[66] Lenells, J. (2005) Traveling wave solutions of the Camassa–Holm equation. *J. Diff. Eq.* **217**, 393–430.

[67] Camassa, R., Holm, D.D. and Hyman, J.M. (1994) A new integrable shallow water equation. *Adv. Appl. Mech.*, **31**, 1–33.

[68] Kalisch, H. and Lenells, J. (2005) Numerical study of traveling-wave solutions for the Camassa–Holm equation. *Chaos, Solitons and Fractals*, **25**, 287–298.

[69] Vakhnenko, V.O. and Parkes, E.J. (2004) Periodic and solitary-wave solutions of the Degasperis–Procesi equation. *Chaos, Solitons and Fractals*, **20**, 1059–1073.

[70] Vakhnenko, V.O. and Parkes, E.J. (2004) The connection of the Degasperis–Procesi equation with the Vakhnenko equation

Proceedings of the Fifth International Conference 'Symmetry in Nonlinear Mathematical Physics', (23-29 June, 2003, Kyiv), eds. A.G. Nikitin, V.M. Boyko, R.O. Popovych, **50**, Part 1, 493–497.

[71] Parkes, E.J. and Vakhnenko, V.O. (2005) Explicit solutions of the Camassa–Holm equation. *Chaos, Solitons and Fractals*, **26**, 1309–1316.

[72] Parkes, E.J. (2008) Periodic and solitary travelling-wave solutions of an extended reduced Ostrovsky equation. *Symmetry, Integrability and Geometry: Methods and Applications, SIGMA*, **4**, no.053, 1–17.

[73] Hone, A.N.W. and Wang, J.P. (2003) Prolongation algebras and Hamiltonian operators for peakon equations. *Inverse Probl.*, **19**, 129–145.

[74] Vakhnenko, V.O. and Parkes, E.J. (1998) The two loop soliton solution of the Vakhnenko equation. *Nonlinearity*, **11**, 1457–64.

[75] Morrison, A.J., Parkes, E.J. and Vakhnenko, V.O. (1999) The N loop soliton solution of the Vakhnenko equation. *Nonlinearity*, **12**, 1427–1437.

[76] Vakhnenko, V.O., Parkes, E.J. and Michtchenko, A.V. (2000) The Vakhnenko equation from the viewpoint of the inverse scattering method for the KdV equation. *Int. J. Diff. Eq. Appl.*, **1**, 429–449.

[77] Estévez, P.G. (2009) Reciprocal transformations for a spectral problem in 2+1 dimensions. *Theor. Math. Phys.*, **159**, 763–9.

[78] Abazari, R. (2010) Application of (G'/G)-expansion method to travelling wave solutions of three nonlinear evolution equation. *Computers and Fluids*, **39**, 1957–1963.

[79] Majid, F., Triki, H., Hayat, T., Aldossary, O.M. and Biswas, A. (2012) Solitary wave solutions of the Vakhnenko–Parkes equation. *Nonlinear Analysis: Modelling and Control*, **17**, 60–66.

[80] Ye, Y., Song, J., Shen, S. and Di, Y. (2012) New coherent structures of the Vakhnenko–Parkes equation. *Results in Physics*, **2**, 170–174.

[81] Roshid, H., Kabir, M.R., Bhowmik, R.C. and Datta, B.K. (2014) Investigation of solitary wave solutions for Vakhnenko–Parkes equation via exp-function and $\exp(-\phi(\xi))$-expansion method. *SpringerPlus*, **3**, 692.

[82] Khan, K. and Akbar, M.A. (2014) The $\exp(-\Phi(\xi))$–expansion method for finding travelling wave solutions of Vakhnenko–Parkes equation. *Int. J. Dynamical Systems and Differential Equations*, **5**, 72–83.

[83] Baskonus, H.M., Bulut, H. and Emir, D.G. (2015) Regarding new complex analytical solutions for the nonlinear partial Vakhnenko–Parkes differential equation via Bernoulli sub-equation function method. *Mathematics Letters*. **1**, 1–9.

[84] Vakhnenko, V.O. and Parkes, E.J. (2002) The calculation of multi-soliton solutions of the Vakhnenko equation by the inverse scattering method. *Chaos, Solitons and Fractals*, **13**, 1819–1826.

[85] Vakhnenko, V.O. and Parkes, E.J. (2012) The singular solutions of a nonlinear evolution equation taking continuous part of the spectral data into account in inverse scattering method. *Chaos, Solitons and Fractals*, **45**, 846–852

[86] Vakhnenko, V.O. and Parkes, E.J. (2012) Solutions Associated with Discrete and Continuous Spectrums in the Inverse Scattering Method for the Vakhnenko–Parkes Equation. *Prog. Theor. Phys.*, **127**, No.4, 593–613.

[87] Vakhnenko, V.O. and Parkes, E.J. (2012) Special singularity function for continuous part of the spectral data in the associated eigenvalue problem for nonlinear equations. *J. Math. Phys.* **53**, No.6, 063504(11).

[88] Vakhnenko, V.O. (2012) Special form of the singularity function for continuous part of the spectral data in inverse scattering method. *Ukr. J. Phys.*, **57**, 95–99.

[89] Vakhnenko, V.O., Parkes, E.J. and Morrison, A.J. (2002) Integrability of a novel nonlinear evolution equation. *Rep. NAS Ukr.*, No.6, 86–92.

[90] Vakhnenko, V.O., Parkes, E.J. and Morrison, A.J. (2003) A Bäcklund transformation and the inverse scattering transform method for the generalised Vakhnenko equation. *Chaos, Solitons and Fractals*, **17**, 683–692.

[91] Vakhnenko, V.O. and Parkes, E.J. (2001) A novel nonlinear evolution equation and its Bäcklund transformation, *Rep. NAS Ukr.*, No.6, 11–16.

[92] Vakhnenko, V.O. and Parkes, E.J. (2001) A novel nonlinear evolution equation integrable by the inverse scattering method. *Rep. NAS Ukr.*, No.7, 81–86.

[93] Vakhnenko, V.O. and Parkes, E.J. (2002) A novel nonlinear evolution equation integrable by the inverse scattering method. *Proceedings of the Fifth International Conference 'Symmetry in Nonlinear Mathematical Physics' (9-15 July, 2001, Kyiv)*, eds. A.G. Nikitin, V.M. Boyko, R.O. Popovych, **43**, Part 1, 384–391.

[94] Gardner, C.S., Greene, J.M., Kruskal, M.D. and Miura, R.M. (1967) Method for solving the Korteweg–de Vries equation. *Phys. Rev. Lett.*, **19**, 1095–1097.

[95] Marchenko, V.A. (1960) Eigenfunction expansions for non-self-adjoint singular differential operators of the second order. *Amer. Math. Soc. Trans.*, **25**, 177–130.

[96] Hirota, R. and Satsuma, J. (1976) A variety of nonlinear network equations generated from the Bäcklund transformation for the Toda lattice. *Suppl. Progr. Theor. Phys.*, No.59, 64–100.

[97] Espinosa, A. and Fujioka, J. (1994) Hydrodynamic foundation and Painlevé analysis of Hirota–Satsuma-type equations. *J. Phys. Soc. Japan*, **63**, 1289–1294.

[98] Ablowitz, M.J., Kaup, D.J., Newell, A.C. and Segur, H. (1974) The inverse scattering transform Fourie analysis for nonlinear problems. *Stud. Appl. Math.*, **53**, 249–315.

[99] Vakhnenko, V.O. and Parkes, E.J. (2006) The solutions of a generalized Degasperis–Procesi equation. *Rep. NAS Ukr.*, No.8, 88–94.

[100] Hirota, R. and Satsuma, J. (1976) N–soliton solutions of model equations for shallow water waves. *J. Phys. Soc. Japan*, **40**, 611–612.

[101] Morrison, A.J. and Parkes, E.J. (2001) The N–soliton solution of a generalised Vakhnenko equation. *Glasgow Math. J.*, **43A**, 65–90.

[102] Morrison, A.J. and Parkes, E.J. (2003) The N-soliton solution of the modified generalised Vakhnenko equation (a new nonlinear evolution equation). *Chaos, Solitons and Fractals*, **16**, 13–26.

[103] Parkes, E.J. (2008) Some periodic and solitary travelling-wave solutions of the short-pulse equation. *Chaos, Solitons and Fractals*, **38**, 154–159.

[104] Schäfer, T. and Wayne, C.E. (2004) Propagation of ultra-short optical pulses in cubic nonlinear media. *Physica D*, **196**, 90–105.

[105] Sakovich, A. and Sakovich, S. (2005) The short pulse equation is integrable. *J. Phys. Soc. Japan*, **74**, 239–241.

[106] Wadati, M., Konno, K. and Ichikawa, Y.H. (1979) New integrable nonlinear evolution equations. *J. Phys. Soc. Japan*, **47**, 1698–1700.

[107] Sakovich, A. and Sakovich, S. (2006) Solitary wave solutions of the short pulse equation. *J. Phys. A: Math. Gen.*, **39**, L361(7).

[108] Turnbull, H.W. (1952) *Theory of Equations*, Edinburgh: Oliver and Boyd.

[109] Boyd, J.P. (1997) Peakons and coshoidal waves: traveling wave solutions of the Camassa–Holm equation. *Applied Mathematics and Computation*, **81**, 173–187.

[110] Abramowitz, M. and Stegun, I.A. (1972) *Handbook of Mathematical Functions*, New York: Dover.

[111] Zabusky, N.J. and Kruskal, M.D. (1965) Interaction of "solitons" in a collisionless plasma and the recurrence of initial states. *Phys. Rev. Lett.*, **15**. 240–243.

[112] *Bäcklund Transformations, the Inverse Scattering Method, Solitons, and Their Applications* (1976), ed. R.M. Miura, New York: Springer.

[113] Rund, H. (1974) Variational problems and Bäcklund transformations associated with the sine-Gordon and Korteweg–de Vries equations and their extensions. *Bäcklund Transformations*, ed. R.M. Miura, Springer-Verlag, New York, pp. 199–226.

[114] Hirota, R. (1974) A new form of Bäcklund transformations and its relation to the inverse scattering problem. *Prog. Theor. Phys.*, **52**, 1498–512.

[115] Satsuma, J. and Kaup, D.J. (1977) A Bäcklund transformation for a higher order Korteweg–de Vries equation. *J. Phys. Soc. Japan*, **43**, 692–697.

[116] Satsuma, J. (1974) Higher conservation laws for the Korteweg–de Vries equation through Bäcklund transformation. *Prog. Theor. Phys.*, **52**, 1396–1397.

[117] Ito, M. (1980) An extension of nonlinear evolution equations of the KdV (MKdV) type to higher orders. *J. Phys. Soc. Japan*, **49**, 771–778.

[118] Hietarinta, J. (1987) A search of bilinear equations passing Hirota's three-soliton condition: I. KdV-type bilinear equations. *J. Math. Phys.*, **28**, 1732–1742.

[119] Hietarinta, J. (1997) Hirota's bilinear method and its generalization. *Int. J. Modern Phys. A*, **12**, 43–51.

[120] Grammaticos, B., Ramani, A. and Hietarinta, J. (1990) A search for integrable bilinear equations: the Painlevé approach. *J. Math. Phys.*, **31**, 2572–2578.

[121] Musette, M., Lambert, F. and Decuyper, J.C. (1987) Soliton and antisoliton resonant interactions. *J. Phys. A: Math. Gen.*, **20**, 6223–6235.

[122] Moloney, T.P. and Hodnett, P.F. (1989) A new perspective on the N-soliton solution of the KdV equation. *Proc. R. Ir. Acad.*, **89A**, No 2, 205–217.

[123] Wazwaz, A.M. (2010) N-soliton solutions for the Vakhnenko equation and its generalized forms. *Physica Scripta*, **82**, 065006(7).

[124] Hodnett, P.F. and Moloney, T.P. (1989) On the structure during interaction of the two-soliton solution of the Korteweg–de Vries equation. *SIAM J. Appl. Math.*. **49**. 1174–1187.

[125] Drazin, P.G. and Johnson, R.S. (1989) *Solitons: an Introduction*, Cambridge University Press.

[126] Wadati, M. (1972) The exact solution of the modified Korteweg–de Vries equation. *J. Phys. Soc. Japan*, **32**, 1681–1681.

[127] Wadati, M. (1973) The modified Korteweg–de Vries equation. *J. Phys. Soc. Japan*, **34**, 1289–1296.

[128] Ablowitz, M.J., Kaup, D.J., Newell, A.C. and Segur, H. (1973) Method for solving the sine-Gordon equation. *Phys. Rev. Lett.*, **30**, 1262–1264.

[129] Flaschka, H. (1974) On the Toda lattice. II. Inverse-scattering solution. *Prog. Theor. Phys.*, **51**, 703–716.

[130] Lax, P.D. (1968) Integrals of nonlinear equations of evolution and solitary waves. *Comm. Pure. Appl. Math.*, **21**, 467–490.

[131] Kaup, D.J. (1980) On the inverse scattering problem for cubic eigenvalue problems of the class $\psi_{xxx} + 6Q\psi_x + 6R\psi = \lambda\psi$. *Stud. Appl. Math.*, **62**, 189–216.

[132] Caudrey, P.J. (1982) The inverse problem for a general $N \times N$ spectral equation. *Physica D*, **6**, 51–66,

[133] Caudrey, P.J. (1980) The inverse problem for the third order equation $u_{xxx} + q(x)u_x + r(x)u = -i\zeta^3 u$. *Phys. Lett. A*, **79**, 264–248.

[134] Zakharov, V.E. (1974) On stochastization of one-dimensional chains of nonlinear oscillators. *Sov. Phys. JETP*, **38**, 108–110.

[135] Deift, P., Tomei, C. and Trubowitz, E. (1982) Inverse scattering and the Boussinesq equation. *Commun. Pure Appl. Math.*, **35**, 567–628.

[136] Musette, M. and Conte, R. (1991) Algorithmic method for deriving Lax pairs from the invariant Painlevé analysis of nonlinear partial differential equations. *J. Math. Phys.*, **32**, 1450–1457.

[137] Clarkson, P. and Mansfield, E.L. (1995) Symmetry reductions and exact solutions of shallow water wave equations. *Acta Appl. Math.*, **39**, 245–276.

[138] Boutet de Monvel, A. and Shepelsky, D. (2013) A Riemann–Hilbert approach for the Degasperis–Procesi equation. *Nonlinearity*, **26**, 2081–2107.

[139] Boutet de Monvel, A. and Shepelsky, D. (2015) The Ostrovsky–Vakhnenko equation by a Riemann–Hilbert approach. *J. Phys. A: Math. Theor.*, **48**, 035204

[140] Caudrey, P.J. (1984) Differential and discrete spectral problem and their inverses. *Wave Phenomena: Modern Theory and Applications*, eds. C. Rogers, T.M. Moodie, Elsevier Science Publishers, North Holland, 221–232.

[141] Pontryagin, L.S. (1962) *Ordinary Differential Equations*, Addison-Wesley Publishing Company, London.

[142] Yusufoglu, E. and Bekir, A. (2008) The tanh and the sine-cosine methods for exact solutions of the MBBM and the Vakhnenko equations. *Chaos, Solitons and Fractals*, **38**, 1126–1133.

[143] Parkes, E.J. (2011) Comment on "Application of (G'/G)-expansion method to travelling wave solutions of three nonlinear evolution equation" [Computers and Fluids 2010;39:1957-1963]. *Computers and Fluids*, **42**, 108–109.

[144] Bao-Feng Feng, Maruno, K. and Ohta, Y. (2012) On the τ-functions of the reduced Ostrovsky equation and the $A_2^{(2)}$ two-dimensional Toda system. *J. Phys. A: Math. Theor.*, **45**, 355203.

[145] Bao-Feng Feng, Maruno, K. and Ohta, Y. (2013) On the τ-functions of the Degasperis–Procesi equation. *J. Phys. A: Math. Theor.*, **46**, 045205.